Wolfgang Schmid

Sputter coating and processing of monolithic U-Mo nuclear fuel

Wolfgang Schmid

Sputter coating and processing of monolithic U-Mo nuclear fuel

Theory, Instrumentation, Application

Südwestdeutscher Verlag für Hochschulschriften

Impressum/Imprint (nur für Deutschland/only for Germany)
Bibliografische Information der Deutschen Nationalbibliothek: Die Deutsche Nationalbibliothek verzeichnet diese Publikation in der Deutschen Nationalbibliografie; detaillierte bibliografische Daten sind im Internet über http://dnb.d-nb.de abrufbar.
Alle in diesem Buch genannten Marken und Produktnamen unterliegen warenzeichen-, marken- oder patentrechtlichem Schutz bzw. sind Warenzeichen oder eingetragene Warenzeichen der jeweiligen Inhaber. Die Wiedergabe von Marken, Produktnamen, Gebrauchsnamen, Handelsnamen, Warenbezeichnungen u.s.w. in diesem Werk berechtigt auch ohne besondere Kennzeichnung nicht zu der Annahme, dass solche Namen im Sinne der Warenzeichen- und Markenschutzgesetzgebung als frei zu betrachten wären und daher von jedermann benutzt werden dürften.

Coverbild: www.ingimage.com

Verlag: Südwestdeutscher Verlag für Hochschulschriften GmbH & Co. KG
Dudweiler Landstr. 99, 66123 Saarbrücken, Deutschland
Telefon +49 681 37 20 271-1, Telefax +49 681 37 20 271-0
Email: info@svh-verlag.de

Approved by: München, TU, Diss., 2011

Herstellung in Deutschland:
Schaltungsdienst Lange o.H.G., Berlin
Books on Demand GmbH, Norderstedt
Reha GmbH, Saarbrücken
Amazon Distribution GmbH, Leipzig
ISBN: 978-3-8381-2594-7

Imprint (only for USA, GB)
Bibliographic information published by the Deutsche Nationalbibliothek: The Deutsche Nationalbibliothek lists this publication in the Deutsche Nationalbibliografie; detailed bibliographic data are available in the Internet at http://dnb.d-nb.de.
Any brand names and product names mentioned in this book are subject to trademark, brand or patent protection and are trademarks or registered trademarks of their respective holders. The use of brand names, product names, common names, trade names, product descriptions etc. even without a particular marking in this works is in no way to be construed to mean that such names may be regarded as unrestricted in respect of trademark and brand protection legislation and could thus be used by anyone.

Cover image: www.ingimage.com

Publisher: Südwestdeutscher Verlag für Hochschulschriften GmbH & Co. KG
Dudweiler Landstr. 99, 66123 Saarbrücken, Germany
Phone +49 681 37 20 271-1, Fax +49 681 37 20 271-0
Email: info@svh-verlag.de

Printed in the U.S.A.
Printed in the U.K. by (see last page)
ISBN: 978-3-8381-2594-7

Copyright © 2011 by the author and Südwestdeutscher Verlag für Hochschulschriften GmbH & Co. KG and licensors
All rights reserved. Saarbrücken 2011

The object of this invention is to produce a coating of one material upon another;...
The uses of the invention are almost infinite, for coatings of any material and of any desired thickness may be formed.

Excerpt from "Art of plating one material with another", the first patent on a sputter deposition process, granted to the inventor Thomas Alva Edison in the year 1894.

Contents

1 **Motivation**   1
   1.1 Background 1
     1.1.1 FRM II 2
     1.1.2 RERTR program 6
     1.1.3 Conversion 9
   1.2 High density fuels 12
     1.2.1 Properties 12
     1.2.2 Assembly 13
     1.2.3 Composition 15
     1.2.4 U-Mo alloys 16
   1.3 Aim of thesis 20

2 **Theory**   21
   2.1 Sputtering 21
     2.1.1 Sputter erosion mechanism 23
     2.1.2 Ejected particles 25
     2.1.3 Target effects 30
     2.1.4 Sputter deposition mechanism 32
     2.1.5 Deposited film 35
     2.1.6 Substrate effects 39
   2.2 Technology 42
     2.2.1 Process 42
     2.2.2 Technical realization 42
     2.2.3 Sputtering setup 45

3 **Instrumentation**   49
   3.1 Construction 49

CONTENTS

	3.1.1	Tabletop reactor	51

        3.1.1 Tabletop reactor 51
        3.1.2 Full size reactor 57
  3.2 Operation 65
        3.2.1 Reactor properties 66
        3.2.2 Target properties 71
        3.2.3 Substrate properties 81
        3.2.4 Processing parameters 83
  3.3 Film growth control 85

4 Application   91
  4.1 U-Mo processing 91
        4.1.1 Film formation 92
        4.1.2 Surface cleaning 101
        4.1.3 Coating 103
  4.2 Application I: Fuel fabrication 107
        4.2.1 Fuel foil 107
        4.2.2 Barrier coating 111
        4.2.3 Cladding 115
        4.2.4 Prospect 120
  4.3 Application II: Scientific samples 122
        4.3.1 Irradiation experiments 123
        4.3.2 Thermal diffusion experiments 130
        4.3.3 Prospect 132

5 Advancement   135
  5.1 Film gradients 135
  5.2 Film pollution 145
  5.3 Target material spectrum 146
  5.4 Target utilization 153

6 Conclusion   157
  6.1 Summary 157
  6.2 Conclusion 158

Appendix   161
  A1 SRIM 161

	A1.1 Program	161
	A1.2 Sputtering yield and mean energy	162
	A1.3 Ion backscattering	163
	A1.4 Atom reflection	164
A2	SPUSI	166
	A2.1 Basic program	166
	A2.2 Implementation of movement	168
	A2.3 Implementation of masks	170
A3	Metallurgy of U-Mo	173
	A3.1 Metallic U	173
	A3.2 U-Mo alloys	173
A4	Tensile tests	176

Bibliography 178

Chapter 1

Motivation

The work presented in this thesis was conducted within the framework of the fuel conversion research program at the research reactor "Forschungsneutronenquelle Heinz Maier-Leibnitz" (or: FRM II) in Garching/Germany. The intention of this program is to clarify, whether there is a possibility to convert FRM II from its current nuclear fuel, that contains highly enriched uranium (^{235}U content of 93%), to a new one that is significantly lower enriched (^{235}U content $\leq$ 50%). The problems related to a fuel conversion like this touch the fields of reactor physics, metallurgy, engineering and processing technology as well as economic aspects and national and international political guidelines.

This first chapter briefly describes the background of the fuel conversion research program, and explains the present need for a high density nuclear fuel. It also locates this thesis into the frame of current international fuel development activities.

1.1 Background

The considerations to build a high flux research reactor at the Technische Universität München (or: TUM) reach back to the 1970s [boe98]. Already before 1980 first design works and calculations for a compact reactor core, that use a fuel containing highly enriched uranium, started in the group of Prof. Gläser. The compact core concept was presented for the first time 1981 at the 'International Conference on Neutron Irradiation Effects' and was first published in 1982 in the 'Journal of Nuclear Materials' [boe82]. In the following years, it was continuously enhanced.

In 1987, first funds were granted by the Bavarian Government, and the so called "Projektgruppe FRM II" ('project group FRM II') could be established that started to work out a detailed physical and technical concept for the construction of a

1 MOTIVATION

new research reactor at TUM. The German "Wissenschaftsrat" ('scientific advisory committee') examined the project in 1989 and concluded its examination with a strong recommendation for the construction of a high flux research reactor in Garching [gla99]. The nuclear licensing process started in 1993 and continued until 2003 when the "3. Teilbetriebsgenehmigung" ('third partial operational license') was finally granted. FRM II was constructed from August 1996 to January 2001 and the first criticality was reached at 2nd of March in 2004 [her04].

1.1.1 FRM II

FRM II combines the technical and regulatory advantages of a research reactor of relatively low thermal power with the neutron flux and usability of a large high power neutron source. It has the purpose to serve as a strong neutron source for science, industry and medicine. Therefore the main aim for the operation of FRM II is to provide a maximum neutron flux to its users for as many days per year as possible.

Construction Regarding the basic design, FRM II is a so called beam tube reactor with a pool. Figure 1.1 shows a cut view of the FRM II pool region. The core of the reactor is a single fuel element in the middle of an Al tank filled with high purity D_2O. The tank is at the bottom of the H_2O containing pool. Huge tubes, filled only with He, penetrate the pool walls and the D_2O tank and are arranged tangentially around the fuel element. These 'beam tubes' allow the neutrons to leave the area close to the core and to reach the experiments, while their tangential arrangement avoids a direct line of sight contact and thus prevents the gamma radiation emitted from the fuel element to reach the experiments.

The fuel element is the primary neutron source in the core region of FRM II. It produces fast neutrons (E $\propto$ 2 MeV) by nuclear fission, that are quickly moderated to thermal energies (E $\propto$ 25 meV) in the D_2O moderator. The D_2O also 'conserves' the neutrons, as it has a very small neutron absorption cross section ($\sigma \approx 1.32$ mbarn). Therefore a high thermal neutron flux of up to $8 \cdot 10^{14} \frac{n}{cm^2 s}$ can build up in the D_2O volume around the fuel element. This flux is extracted via the beam tubes but also used to supply several secondary neutron sources as the hot neutron source (for epithermal neutrons, E $\propto$ 10 eV), the cold neutron source (for cold neutrons, E $\propto$ 10 meV), the ultra-cold source (for ultra-cold neutrons, E $\propto$ 100 neV) or the converter plate (for fast neutrons, E $\propto$ 2 MeV). With its primary and its secondary sources FRM II can thus provide a wide spectrum of epithermal, thermal and subthermal neutrons for wide variety of applications. Figure

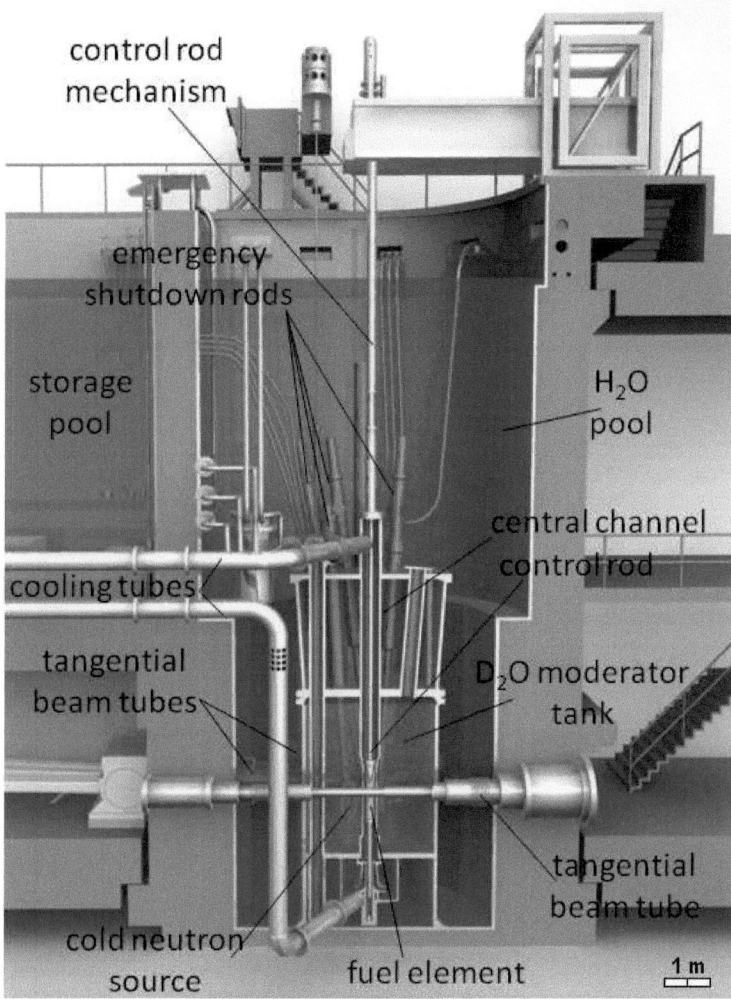

Figure 1.1: Cut view of the FRM II pool region (from [for09]). The D_2O moderator tank is positioned at the bottom of the H_2O pool. During operation the fuel element is mounted inside a vertical channel that leads centrally through the D_2O tank and is part of the cooling circuit. The D_2O tank houses also the secondary neutron sources. Beam tubes filled with He represent openings in the D_2O volume, that also works as a reflector, and allow the neutrons from primary and secondary sources to leave the core area and to reach the experiments. The H_2O pool outside has primarily no meaning for reactor physics but serves as a biological shielding. Next to the H_2O pool is a storage pool for spent fuel elements.

1 MOTIVATION

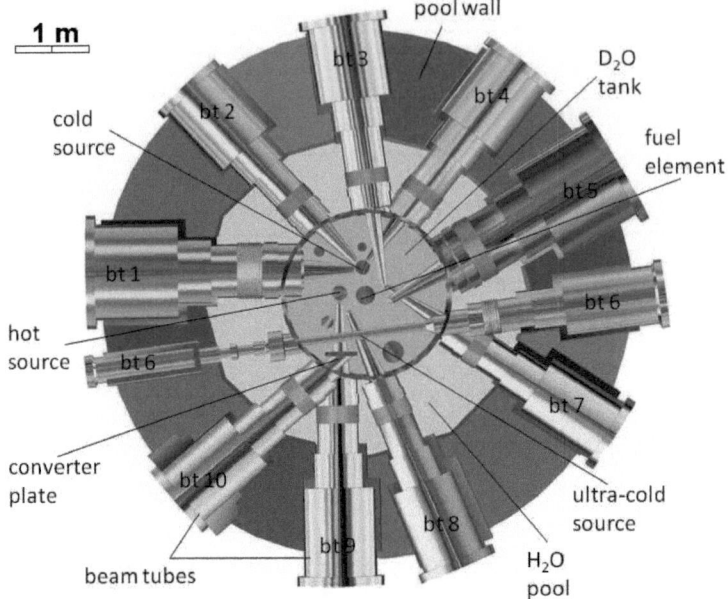

Figure 1.2: Top view of the FRM II core region with its concentric structure (from [for09]). The closed D_2O moderator tank is located at the bottom of the H_2O containing pool. The pool itself is limited by the pool wall. A vertical tube leads through the center of the D_2O tank. The fuel element and the control rod are positioned inside this chimney. Inside the D_2O there are four secondary neutron moderators (called 'hot source', 'cold source', 'ultra-cold source' and 'converter plate') located close to the fuel element as well as a positron source and several irradiation positions. Ten horizontal beam tubes (denoted as bt 1 - 10) are arranged tangentially and allow neutrons from the primary and secondary sources to leave the core region and reach the experiments.

1.2 shows a top view of the core region with the ten beam tubes as well as the primary and secondary neutron sources.

Compact core   The design of the FRM II reactor core is based on a compact core concept that was developed at TUM [boe82]. The core contains only one single fuel element at the center of a moderator tank filled with high purity[1] D_2O (see figure 1.3).
A stream of H_2O flows through channels inside the fuel element and guarantees a permanent cooling during operation. Both the D_2O in the moderator tank around

---

[1]D_2O is usually contaminated by H_2O. In FRM II the H_2O impurity in the D_2O has to be $\leq$ 0.2 %. Otherwise the neutron absorption in the tank will be too large to allow a criticality in the reactor.

Background

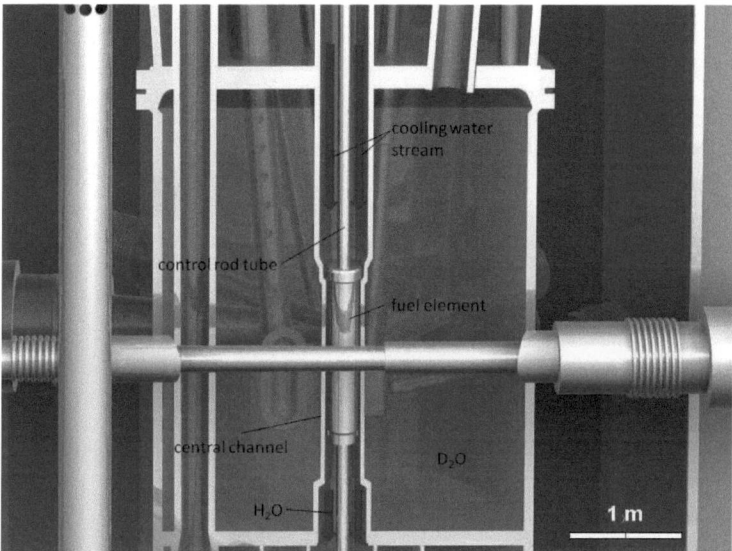

Figure 1.3: Cut view of the FRM II core region (from [for09]). The fuel element is positioned inside the central channel. Inside the fuel element there is the control rod that is used to regulate reactor power. During reactor operation the fuel element produces a thermal power of 20 MW. A major fraction of this power (about 19 MW) is removed by the primary cooling circuit, i.e. cooling water (H_2O) from the primary circuit streams from the top side through the fuel element and back to the system. The remaining power is deposited into the D_2O moderator and transported into the pool by natural convection. The D_2O moderator tank is however a closed system and completely autarkic from the primary cooling circuit and the pool.

the element and the H_2O inside the cooling channels of the element serve as neutron moderators. The D_2O further has the function of a neutron reflector for the fuel element. The whole arrangement of fuel, surrounding D_2O and internal H_2O allows to reach criticality even tough the fuel element is very compact and its uranium load is with 8.1 kg (7.5 kg ^{235}U) small compared to other high flux reactors[2]. The compactness on the other side provides inherent safety features as well as a very high thermal neutron flux of up to $8 \cdot 10^{14} \frac{n}{cm^2 s}$ at a reactor power of only 20 MW_{th}[3]. Because of the special compact design FRM II is currently the reactor with the highest ratio of neutron flux to reactor power in the world [for09].

---

[2]High Flux Isotope Reactor (HFIR, Oak Ridge National Lab): uranium load 10.1 kg (9.4 kg ^{235}U) [xou04], Réacteur Haut Flux (RHF, Institut Laue Langevin): uranium load 9.2 kg (8.6 kg ^{235}U) [mo89].

[3]HFIR: $2.6 \cdot 10^{15} \frac{n}{cm^2 s}$ @ 85 MW [xou04], RHF: $1.5 \cdot 10^{15} \frac{n}{cm^2 s}$ @ 58 MW [rap09].

1 MOTIVATION

Fuel element The fuel element of FRM II is an upright standing cylindrical arrangement of 113 single fuel plates inside a cylindrical tube (see figure 1.4).

All fuel plates are involutely bended along their short side. Aluminum spacers and combs keep the plates in position and guarantee a constant gap of 2.2 mm between each pair of plates for an equal distribution of cooling water stream.

A control rod of metallic Hf with a Be follower is located and moved inside the cylindric central channel of the fuel element. At the beginning of a reactor cycle it is nearly completely inside the fuel element and during reactor operation it is moved more and more out of the central channel as the burnup[4] of the fuel increases. A B ring, located at the bottom of the fuel element, has the function to avoid the formation of undesired large temperature gradients along the height of the element.

Every fuel element can be operated for a total of 60 days at a maximum thermal power of 20 MW, which is denoted as one reactor cycle, until the control rod is completely drawn out from the central channel. At this point the excess reactivity[5] of the fuel element approaches zero and it has to be replaced. The fuel has then reached an average ^{235}U burnup of 20.4% [bre11].

Fuel FRM II uses the intermetallic compound U_3Si_2 as a nuclear fuel. The material contains U with an ^{235}U isotope content of 93 at%, which is the fuels' primary fissile isotope [nud00]. A grinded powder of the U_3Si_2 fuel is embedded into an Al matrix forming a so called U_3Si_2-Al 'meat' structure. There are two coherent meat zones in a fuel plate (see figure 1.5): an inner zone with a U_3Si_2/Al volume ratio of 27 vol%[6] and an outer zone with a U_3Si_2/Al volume ratio of 14 vol%. The meat zones have a size of together 700 mm x 62.4 mm and a thickness of 600 μm. They are completely surrounded by a so called 'cladding' layer consisting of the alloy AlFeNi[7] with 380 μm thickness (see figure 1.5). The final structure, the fuel plate, has a size of 720 mm x 76 mm and a thickness of 1360 μm [har04].

1.1.2 RERTR program

The 'Reduced Enrichment for Research and Test Reactors' (or: RERTR) program was launched in 1978 by the US-Department of Energy (DOE). The aim of the program is to provide the fuel technology and the analytical support required

---

[4] The percentage of the initial fissile isotope inventory that has been consumed by fission is denoted as 'burnup'.

[5] Excess reactivity denotes the amount of surplus reactivity over that needed to achieve criticality in the reactor. Excess reactivity is brought into a reactor by the installation of additional fuel in the core and is supposed to compensate fuel burnup and the buildup of neutron poisons.

[6] Also called fuel volume loading, see substection 1.2.2.

[7] See table 3.3 in chapter 3 for the exact composition of this Al alloy

Background

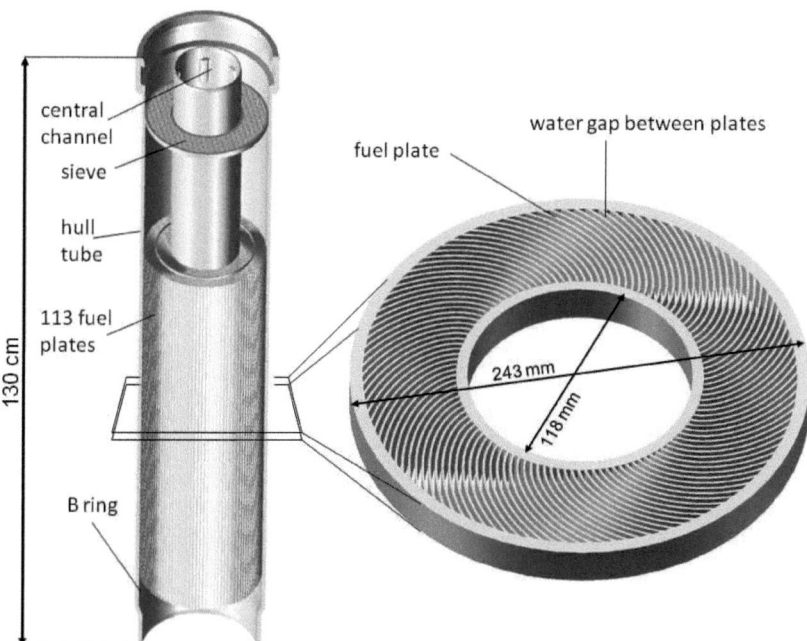

Figure 1.4: Cut of the FRM II fuel element (fuel element model from [bre11]). The fuel is contained in 113 single fuel plates that are mounted between an inner and an outer tube. Between each two fuel plates there is an equidistant gap that allows cooling water to flow between the plates. A sieve that is located above the fuel plates stops particles that are larger than the gap between the plates and could possibly block the cooling channels. The inner tube defines the central channel where the control rod is located during reactor operation. The outer tube is the hull tube that reinforces the structure. On the bottom a B ring surrounds the fuel element. B is a potent neutron poison (σ_{abs} = 764 barn), and the ring has the effect of suppressing the neutron flux in the bottom area of the fuel element. This limits the local fission power density at the beginning of the cycle and thus avoids local hot spots.

1 MOTIVATION

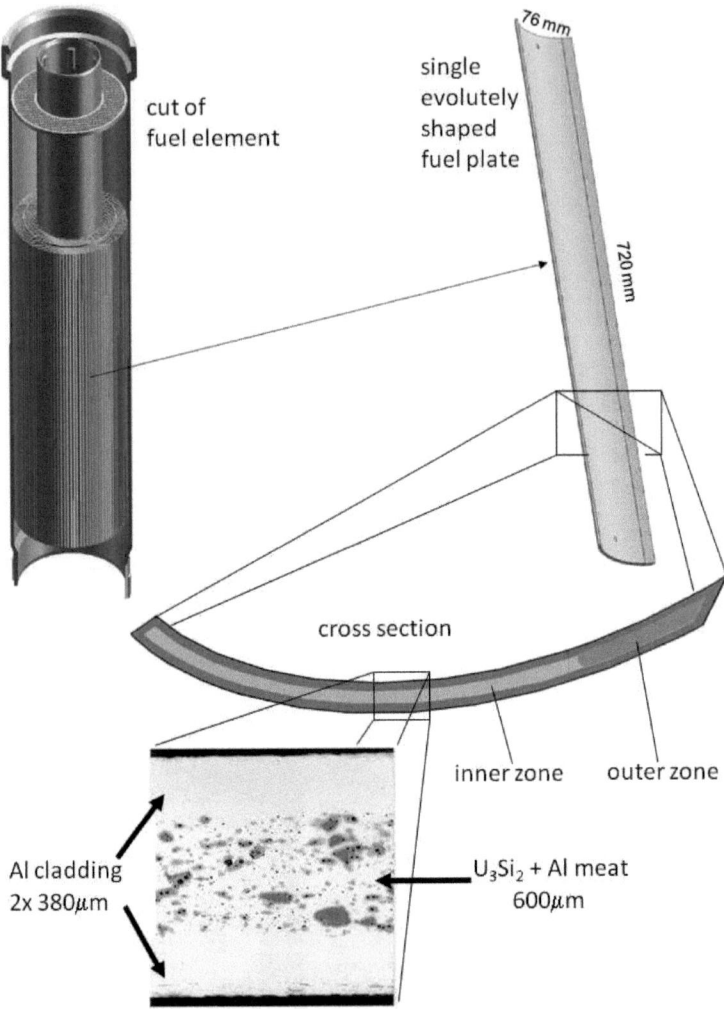

Figure 1.5: The fuel element is an arrangement of 113 involutely shaped fuel plates (fuel element model from [bre11]). Each fuel plate has a size of 720 mm x 76 mm and a thickness of 1360 μm. It consists of a meat kernel that is surrounded by an AlFeNi cladding layer of 380 μm thickness. The meat itself is composed of U_3Si_2 fuel powder dispersed into an Al matrix. The U_3Si_2/Al volume ratio is 27 vol% for the inner meat zone and 14 vol% for the outer meat zone.

Background

to convert research and test reactors worldwide from nuclear fuels that utilize highly enriched uranium (HEU[8]) to fuels based on low enriched uranium (LEU[9]) [rob09].
Since 1978 about 40 reactors could be converted to LEU by using traditional fuel types including U_3Si_2 [rertr]. Anyhow, some research reactors were identified, that cannot be converted by this fuel types for several reasons, and FRM II is one of these reactors. A conversion of these facilities would need novel fuel materials with a very high uranium density.

Obligation Today FRM II is the only research reactor in Germany that uses nuclear fuel containing HEU. As the operator of FRM II, TUM has engaged itself to take all necessary measures to reduce the uranium enrichment of the used fuel to ≤ 50 at% if it is technically feasible without compromising the operation of the facility [tg3]. TUM is therefore participating in the RERTR program and also established its own fuel conversion research program [boe04]. The aim of this program is an investigation of possible high-density nuclear fuels, that could allow a conversion of FRM II (see subsequent section), and a study of the metallurgy and the behavior of these materials under the irradiation conditions given inside a reactor core. However, the program aims as well towards a near-term utilization of these fuel materials, i.e. its activities also involve the development of techniques for high-density fuel plate fabrication.

1.1.3 Conversion

Different conversion scenarios for FRM II were investigated by the fuel conversion research program [boe02],[boe04]. A prerequisite for all of these scenarios was a conservation of the very high safety standard and the scientific quality of the neutron source [boe04].

Conversion considerations A first main result gained by the fuel conversion research program was the conclusion, that any realistic and technically feasible conversion scenario for FRM II could only concern the nuclear fuel itself, as the size and geometry of the fuel element and the reactor power should retain unchanged [boe02],[boe04]. Alternative scenarios, that are based on a change of size or geometry of the fuel element, or that try to increase reactor power, would require a major modification of core components or reactor systems and therefore generate enormous costs and a long reactor downtime, which both should be avoided.

---

[8] According to IAEA defined as uranium with ≥ 20 at% ^{235}U isotope content.
[9] According to IAEA defined as uranium with ≤ 20 at% ^{235}U isotope content.

1 MOTIVATION

As already mentioned, the aim for the operation of FRM II is to provide its users a maximum neutron flux at a maximum cycle length. To preserve both parameters at a reduced level of fuel enrichment means, that the ^{235}U inventory in the core has to stay constant. In fact, it has even to be increased, as a lower fuel enrichment also means a higher concentration of the neutron absorbing ^{238}U, and the additional absorption has consequently to be compensated by an additional increase of the ^{235}U inventory. Therefore, the fuel element used after conversion would indeed have a lower enrichment of ^{235}U, but at the same time it would have to contain a much larger amount of ^{235}U and therefore a much larger total amount of U than the present fuel element.
Both factors together, an increased ^{235}U inventory at a lower enrichment and a fixed size and geometry of the fuel element, lead to the conclusion, that the fuel material in a converted fuel element has to provide a much higher U density than the U_3Si_2 that is used today, as the additional U inventory can only be stored in the given geometry by increasing the fuel density.

Conversion scenario Röhrmoser calculated the necessary amounts of U and ^{235}U as well as the necessary U density of a fuel material for the given fuel element geometry and the given prerequisites [roe05]. Figure 1.6 illustrates the results graphically. As already mentioned, the inventories of ^{235}U and U have to be increased more than linearly with decreased enrichment due to absorption in ^{238}U. As the fuel volume is fixed, the fuels' U density has to increase more than linearly just as well. An upper limit for the U density and likewise for the U and ^{235}U inventory in the fuel element is given by the density of pure metallic U in the alpha phase with 19.06 $\frac{gU}{cm^3}$ [hol58], as no material can have a larger U density. This natural U density maximum defines the theoretical lowest achievable enrichment for a FRM II fuel element with about 31 at%^{235}U, which is just one third of the present enrichment and far below the 50 at% value that was obligated for FRM II by contract, but is still above the maximum enrichment of LEU. Figure 1.6 thus clearly shows, that it is not possible to convert FRM II to a fuel based on LEU under given circumstances (see also [bre11]). It further shows, that any realistically feasible conversion scenario for FRM II requires a HEU fuel with reduced enrichment (which is often referred to as 'medium enriched uranium' or 'MEU') but increased U density.
The dispersed U_3Si_2 fuel used today in FRM II has a maximum U density of up to 3.0 $\frac{gU}{cm^3}$ with an ^{235}U content of 93 at%. To reach an enrichment value of 50 at%, about 8.0 $\frac{gU}{cm^3}$ will be necessary. So theoretically a nuclear fuel needs to have a U density between 8.0 - 19.06 $\frac{gU}{cm^3}$ and would by that allow an enrichment of 31 - 50 at%^{235}U in a conversion scenario. Unfortunately, at the moment there is no

Background

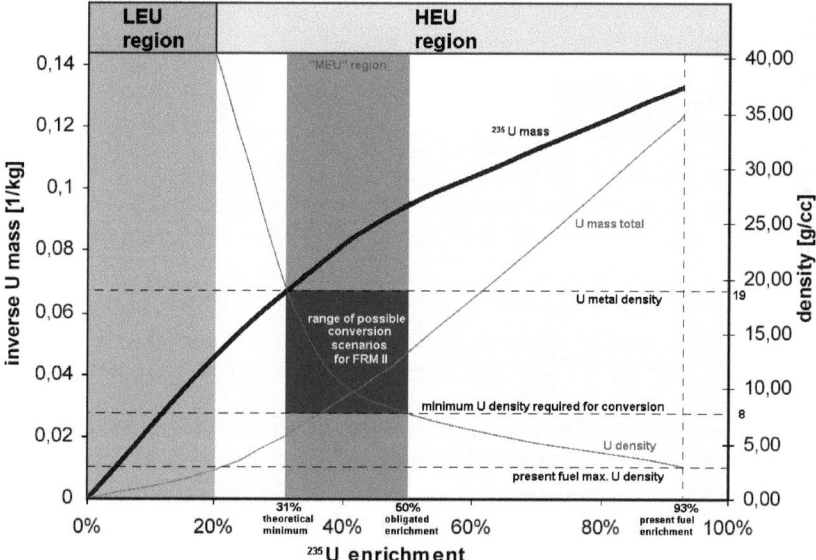

Figure 1.6: Correlation between required ^{235}U mass, total U mass, U density and ^{235}U enrichment of a fuel assuming the present geometry of the FRM II fuel element (according to [roe05]). The masses are given as inverse values, as lower enrichment of course means larger inventories of both U and ^{235}U. The inventory of ^{235}U increases more than linearly with decreasing enrichment, as smaller fraction of ^{235}U means a higher fraction of ^{238}U and thus a higher absorption. The U density of the fuel is determined by the total U inventory and the fixed fuel volume in the fuel element. A 'conversion window' results, when the minimum required U density of 8 $\frac{gU}{cm^3}$ and the maximum possible U density of 19 $\frac{gU}{cm^3}$ are considered. The mass, density and enrichment parameters of all feasible conversion scenarios are limited to the values covered by this window.

11

1 MOTIVATION

qualified high-density nuclear fuel that can provide an U density of this order of magnitude.

1.2 High density fuels

To identify an appropriate high-density nuclear fuel material, the necessary properties have to to be specified. The properties themselves together with the way of assembly then specify the composition of the fuel.

1.2.1 Properties

The technically relevant properties that allow to use a material as a nuclear fuel are its fissile isotopes density, its thermal conductivity as well as its expansion behavior under the influence of heat and radiation (also referred as dimensional stability).

Fissile isotopes density The technically relevant fissile isotopes are ^{233}U, ^{235}U, ^{239}Pu and ^{241}Pu. Only ^{235}U will be focused here as fissile isotope, as it is the primary fissile isotope in the fuel element of FRM II.
It was shown in the previous section, that fuel with a U density of at least 8.0 $\frac{gU}{cm^3}$ will be necessary to convert FRM II to ≤ 50 at%^{235}U enrichment. This value can be seen as a fixed lower limit for our further considerations. The upper limit of U density is not fixed but given by the particular composition of the fuel that will be used. Pure metallic U would provide the maximum possible U density of 19.06 $\frac{gU}{cm^3}$, but is out of question as it shows an instable behavior under reactor conditions (see also appendix A3). The addition of certain materials to U can stabilize its behavior, but automatically reduces the U density of the fuel. Therefore 19.06 $\frac{gU}{cm^3}$ can somehow be seen as the natural upper limit for U density, but realistic U densities will be well below this value.

Thermal conductivity The thermal conductivity of the U_3Si_2-Al fuel (U_3Si_2 dispersed in Al) currently used in FRM II is in between the thermal conductivities of pure Al (235 $\frac{W}{m \cdot K}$) and pure U_3Si_2 (15 $\frac{W}{m \cdot K}$) and determined by the degree of dispersion. Dependent on the fuel volume loading, it has a value of about 109 $\frac{W}{m \cdot K}$ for 27 vol% U_3Si_2/Al volume ratio respectively about 165 $\frac{W}{m \cdot K}$ for 14 vol% U_3Si_2/Al volume ratio. The temperatures reached thereby in the fuel are approximately 130 °C [bre11]. These parameters are not fixed however and can be varied within reasonable limits when converting to a new fuel material. Moreover it is hardly possible to give a defined lower limit of thermal conductivity before the fuel selection, as this depends strongly on the composition of fuel material used and the temperatures it can tolerate.

Dimensional stability In contrast to the fissile isotope density and thermal conductivity, that immediately affect the behavior of the reactor, the dimensional stability of the fuel gets important when the reactor is operated for longer times. It is self-evident that major deformations of reactor core components are in general undesired as it always imposes the danger of malfunctions, damages or even accidents. Nevertheless, a certain amount of deformation, the so called 'swelling', will appear in fuel elements and cannot be avoided. It is possible already during the construction of a fuel element to account for a certain amount of swelling and thus to control this effect. If the swelling is however larger than the accounted amount, the undesired potentially dangerous situation is given again. It is therefore necessary to understand the swelling effect in fuel elements and to minimize it.

The swelling of fuel elements is in general always a result of a change in the geometry of fuel and cladding materials during reactor operation. The reason for this change are heat and radiation driven effects that include phase transformations, diffusion and diffusion induced chemical reactions between different materials, the accumulation of fission products in the fuel matrix. To reduce the swelling to a minimum amount, these effects have to be reduced or avoided if possible.

For the FRM II fuel element the swelling effect is especially critical, as a thickness increase of the fuel plates narrows the width of the cooling channels in between the single plates. Swelling directly lowers therefore the cooling water flux through the fuel element and thus increases the fuel temperature during operation. The maximum acceptable fuel temperature sets here a limit for an acceptable amount of controlled swelling. The appearance of uncontrolled swelling has to be avoided completely.

1.2.2 Assembly

There are two common types of fuel assembly for research reactors, plate type and pin type. Only the plate type should be considered here, as it allows a higher power density and is therefore used by all high flux reactors worldwide.

In plate type fuel elements, the fuel is contained in plates or blank sheets that can be flat or bent. The plates consist of a central meat region as well as a cladding around the meat. There are two existing designs of meat: so called monolithic meat, that consists only of pure fuel material, and so called dispersed meat, where the U-Mo fuel material is distributed inside a matrix material. Generally the meat layer is always completely surrounded by a cladding, which is the only barrier

1 MOTIVATION

between meat and cooling water.

Dispersed meat Dispersed meat is formed by a fuel powder which is embedded into a carrier matrix, so that nearly every fuel powder grain is surrounded by the matrix material. This design provides a large contact area between fuel and carrier matrix and guarantees a good heat removal from the single fuel grains. Due to this the dispersed meat can be operated at high power. However it also offers a large interaction area for reactions between fuel and matrix material and for heat and radiation driven atomic migration between both.
Dispersed meat was investigated since the 1950s [fue57] and the existing technologies for manufacturing it are simple and well developed. A drawback of dispersed meat is the fact, that with increasing fuel to matrix volume ratio (or: fuel volume loading) the thermal conductivity of the dispersed meat decreases drastically [cah94]. Furthermore from manufactors experience a fuel volume loading of 55 vol% seems to be the upper limit reachable for a commercially viable process [sne96].

Monolithic fuel Increasing the fuel volume loading of the meat to 100 vol% one reaches the so called monolithic fuel. Here the meat is one single massive fuel layer that is surrounded directly by the cladding. This design offers the advantage of a small interaction area between fuel and cladding and the maximum achievable fuel volume fraction [cla03]. Unfortunately, it has also the disadvantage that the heat transport from the fuel into the cladding and further into the cooling water is mainly dependent and limited by the thermal conductivity of the fuel.
At the moment, the technologies for manufacturing monolithic fuel are in development.

Cladding The cladding fulfills two very important functions in the fuel element: first, it provides the thermal contact between meat and cooling water that is needed to remove the fission heat from the fuel. A good thermal contact between meat and cladding guarantees a steady and predictable transport of the fission heat from the fuel into the cooling water. If this contact worsens[10], a steady heat removal from the fuel cannot be guaranteed any more and the fuel temperature rises to some new equilibrium temperature or, in the worst case, even to the fuels melting point. Second, it separates the fuel and the fission products that accumulate during reactor operation from the cooling water outside. A separation of the fuel from water is mandatory, as otherwise the hot unprotected fuel would face

---

[10]The two reasons for a worsening of the thermal contact are a delamination of meat and cladding or the buildup of new phases or chemical compounds between fuel and cladding.

High density fuels

compound	U density $[gU/cm^3]$	compound	U density $[gU/cm^3]$
U_3Si_2	11.3	U_2Tc	13.9
UB_2	11.6	U_2Ru	13.9
UCo	12.3	U_3Si	14.6
UC	13.0	U_6Co	17.0
UN	13.5	U_6Ni	16.9
U_2Ti	13.7	U_6Fe	17.0
U_2Mo	13.8	U_6Mn	17.1

Table 1.1: Some high-density uranium compounds that were identified during the US screening campaign in the 1990s (according to [sne96]). In subsequent investigations it was found, that none of them is suited to be used as high-density nuclear fuel.

a massive chemical erosion in the quick stream of cooling water. A retainment of the fission products within the cladding is also desirable, as otherwise the cooling water and the cooling system of the reactor would be massively contaminated.
Both functions are of outstanding importance for the operation and safety of the reactor. Therefore the integrity of the cladding and a stable thermal contact between fuel and cladding have to be guaranteed at any time.

1.2.3 Composition

From the view of reactor physics, it is usually not necessary to contain fissile isotopes into fuel alloys or compounds, as only the types of present nuclides, their cross sections for interactions with neutrons and their three-dimensional arrangement are of relevance. From the view of engineering, materials science and solid state physics, the use of such compounds or alloys is however mandatory most of the time to construct a realistically and safely working fuel element.

Material selection Mid of the 1990s, working groups at the Argonne National Laboratory (or: ANL) and later at the Idaho National Laboratory (or: INL) started an effort to develop dispersed fuels with an U density of 8 - 9 $\frac{gU}{cm^3}$ for the RERTR program [sne96]. They screened the relevant literature to identify all known uranium alloys and compounds that offered available material data and had a density greater than U_3Si_2. Tables 1.1 and 1.2 show a selection of some of the alloys and compounds that were identified during this screening.

From the list of identified fuel materials only those with a bulk U density of at least 15 $\frac{gU}{cm^3}$ were considered for further investigations. This value comes from

1   MOTIVATION

alloy	U density [gU/cm^3]	alloy	U density [gU/cm^3]
U-4Mo	17.4	U-6Mo-0.6Ru	16.5
U-5Mo	17.0	U-6Mo-1Pt	16.5
U-8Mo	16.0	U-4Zr-6Nb	14.8
U-9Mo	15.5	U-4Zr-2Nb	16.3
U-10Mo	15.3	U-3Zr-9Nb	14.3
U-10Mo-0.5Sn	15.3	U-3Zr-5Nb	15.5

Table 1.2: Some high-density uranium alloys that were identified during the US screening campaign in the 1990s (according to [sne96],[hof99]). The RERTR irradiation scoping campaigns revealed, that binary and ternary uranium molybdenum (U-Mo) alloys with 6 - 10 wt% Mo content show an excellent irradiation performance and could be suited to be used as high-density nuclear fuels.

the fact, that the U density of a dispersed fuel is given as the bulk U density of the pure fuel material times the fuel volume loading in the dispersion. Accounting the maximum produceable fuel volume loading today, which is 55 vol% [sne96], a bulk U density of at least 15 $\frac{gU}{cm^3}$ would thus be necessary to reach a dispersed U density of 8 - 9 $\frac{gU}{cm^3}$. The remaining materials consisted of U alloys with small amounts of other metals and of intermetallic U compounds of the structure U_6X (like U_6Fe or U_6Mn).

The selected high-density fuel materials were reviewed concerning their irradiation behavior. Several irradiation tests indicated, that all U_6X compounds could be subject to so-called breakaway swelling[11] or show other poor swelling properties [hof87], thus they were excluded from further study.

Dispersed samples of the remaining U alloys were tested in two low temperature irradiation scoping campaigns at the ATR reactor at INL called RERTR-1 and RERTR-2. Post-irradiation examinations (or: PIEs) of the samples revealed, that binary and ternary alloys of U with 6 - 10 wt% Mo content show an excellent irradiation performance up to high burnups while all other alloys performed poorly [sne99],[mey02]. Subsequent studies were therefore focused mainly on U-Mo alloys.

1.2.4   U-Mo alloys

A series of further irradiation campaigns followed to study the 'in-pile behavior'[12] of dispersed U-Mo alloys at high temperatures, heat fluxes and burn-ups

---
[11]The term breakaway swelling denotes a swelling that has led to a material volume increase of $\geq$ 100% [que98].
[12]The term 'in-pile' denotes the conditions inside a reactor core during reactor operation.

High density fuels

(see table 1.3). These tests were necessary steps on the way to qualify these materials as nuclear fuels.

IDL formation Although RERTR-1 and RERTR-2 had demonstrated very promising results, later irradiation campaigns revealed an unexpected problem. At elevated temperatures, an irradiation supported diffusion process could be observed between the dispersed U-Mo and the Al matrix, that leads to the formation of a so-called interaction diffusion layer (or: IDL) between both materials [mey00]. The fuel/matrix interaction product showed a very bad thermal conductivity that worsened the heat contact between fuel and cladding significantly. Moreover, it was subject to porosity formation in the meat resulting in an increased swelling behavior, and was prone to the formation of large fission gas bubbles that eventually lead to the onset of breakaway swelling and meat/cladding delamination at higher burn-ups [mey00],[ham05],[wac08]. The worldwide research activities thus focused on the topic IDL formation and prevention.

IDL prevention A first and very simple idea to reduce IDL formation was the minimization of contact area between U-Mo and Al. The so-called monolithic fuel design, where the U-Mo fuel is no longer dispersed as small grains in the Al but only present as one compact structure, is based on that idea. Several irradiation campaigns showed the effectivity of this measure (see table 1.3). Unfortunately, the monolithic design cannot completely avoid IDL formation and has the disadvantage of reduced thermal conductivity.
Another idea that is also under investigation is to apply certain additives to the U-Mo fuel or to the Al matrix that slow down IDL formation. This measure was proposed as a way to form interaction products more similar to the stable materials observed for example in U_3Si_2 based dispersion fuels. The addition of Ti, Zr, V, Nb, Pt, Bi, Mg and Si to the fuel as well as to the Al matrix was investigated extensively and first results are promising [hof06],[par05] [kim05],[jun11].
A way not only to reduce but to completely avoid IDL formation could be to replace Al in the fuel system. For the dispersed fuel design it was attempted to replace the Al matrix by a Mg matrix, as Mg does not react with U. However, it turned out that the manufacturing of a Mg matrix fuel is extremely difficult due to matrix/cladding interactions during fabrication [dub06]. Similar problems are expected for other alternative matrix materials as well. Therefore the topic was not further studied and other alternative matrix materials have not been investigated up to now. For the monolithic fuel design a replacement of the Al cladding by Zr cladding was studied[13] [ari10]. The measure had the desired effect and

---

[13]More precisely the Zr alloy Zry-4 was used, see table 3.3 for the composition.

1 MOTIVATION

campaign title	U-Mo sample description	essential result
RERTR-3	dispersed nanoplates	Extensive IDL formation between U-Mo alloys and Al matrix at elevated temperatures and medium burn-ups [mey00].
IRIS-1 UMUS FUTURE-1 IRIS-2	dispersed full-size plates	Excellent behavior at low temperatures. Extensive IDL formation and swelling at elevated temperatures. Porosity formation and breakaway swelling for medium burn-ups [ham05].
RERTR-4 RERTR-5	dispersed, monolithic miniplates	IDL formation between U-Mo alloys and Al matrix strongly temperature dependent. Monolithic samples show excellent behavior. Si addition seems to reduce IDL [hof04].
RERTR-6 RERTR-7 RERTR-8 RERTR-9	dispersed, monolithic miniplates matrix Si addition	Si addition of $\geq 2\%$ reduces IDL formation. Monolithic U-Mo shows thin IDL between fuel and cladding [wac08].
AFIP-1 IRIS-3 IRIS-TUM	dispersed full-size plates matrix Si addition	Si addition reduces IDL formation and porosity [rob10].
AFIP-2 AFIP-3	monolithic full-size plates Si and Zr barriers	Good performance, no delamination, formation of thin IDL [rob10].
IRIS-4	dispersed full-size plates matrix Si addition oxide coated particles	Si addition reduces IDL formation. Only minor influence of oxide coating [rip09]

Table 1.3: Brief review of the most important irradiation campaigns worldwide to test the in-pile performance of U-Mo alloys. Descriptions and results are very much simplified. For more details see the cited publications.

High density fuels

no IDL formation could be observed during irradiation. The low thermal conductivity of Zr (22.6 $\frac{W}{m \cdot K}$) is however considered as a fundamental problem of Zr cladding, and thus was not accepted as satisfying solution to the problem.

The most promising idea today to avoid IDL formation in both the dispersed as well as monolithic fuel design is the separation of U-Mo and Al from each other by a barrier of an appropriate third material. For dispersed meat, a coating should surround each single fuel grain inside the matrix, for monolithic meat it should surround the monolithic fuel core. An appropriate coating material is supposed stop radiation enhanced thermal diffusion reactions by replacing Al in the diffusion process, either by not forming reaction products with U-Mo at all or by forming only stable reaction products.

Barrier application Several methods to apply diffusion preventive barrier coatings onto dispersed and monolithic U-Mo fuel have been investigated in the last years. Most of them work however only for certain barrier materials.

Oxide coatings for dispersed fuel grains were produced simply by the oxidation of hot U-Mo fuel during air contact. It has been shown however, that coatings from UO_2 have only a small effect on IDL formation [rip09],[jun11]. A similar nitridation reaction can be used to create UN_X coatings, if hot fuel is exposed to a pure nitrogen atmosphere. KAERI is currently testing UN_X coatings created by this method, the irradiation behavior of UN_X hasn´t been studied yet.

Pasqualini proposed chemical vapor deposition (CVD) to coat U-Mo fuel [pas04]. He used a chemical reaction based on silane to condense Si on the surface of U-Mo fuel powder. The same method seems to be feasible as well for monolithic fuel. Si has proven to decrease IDL formation by the formation of U_3Si_2. Silanes are however known to be dangerous in handling, as these substances are extremely toxic and highly reactive. Pasqualini also describes dipping and painting as methods to coat monolithic U-Mo fuel foils [pas04]. The foils are dipped into or painted with appropriate chemicals to form Mg and Ge coatings. The effect of coatings of this type has never been tested however.

Plasma spraying has been used to apply Si coatings [moo08] as well as ZrN [izh09] onto monolithic U-Mo fuel. Coatings of ZrN have shown to strongly decrease IDL formation during irradiation.

The current standard process to coat monolithic U-Mo fuel is colamination during U-Mo foil fabrication [moo08]. Zr coatings of 20 μm thickness have proven to nearly prevent IDL formation. The co-rolling process is however extensive and error-prone [tec08].

A technique, that is well known and widely used for the application of coatings is sputter deposition. In contrast to all of the mentioned techniques, sputter deposition could be used to apply any type of coating onto U-Mo fuel. This promising method is thus investigated in the framework of the FRM II fuel con-

1   MOTIVATION

version research program. Currently groups at FRM II [ste11] and at SCK/CEN in Mol/Belgium [van10] investigate the process for the dispersed fuel design. The application of the process for the monolithic fuel design, which is also the major topic of this thesis, is currently only investigated by a group at FRM II [jar09].

1.3   Aim of thesis

The basic idea to use ion sputter erosion and deposition on monolithic U-Mo fuel was developed at TUM already in 2006 [pat06]. The technique provides the possibility of coating, cleaning and processing of the material, and thus seemed to have several possible applications in the field of fuel fabrication. This thesis represents the first detailed investigation of this topic. The aim of the thesis was therefore to realize an experimental setup, that allows to perform both sputter erosion and deposition on full-sized monolithic U-Mo fuel, and to investigate a potential application of the process in fuel fabrication.

Chapter 2

Theory

It is not known, who was the first one to discover the material erosion and deposition caused by a gas discharge, but the effect that is known today as 'sputtering' was first studied by Grove in 1852 [mat03]. It took until the beginning of the 20th century until the sputtering effect was used commercially for the first time by Thomas Alpha Edison[1].

Today, sputtering is a common coating technique in industry and science. Various different realizations and applications of it have been studied and the sputtering behavior of basically every pure chemical element as well as a large variety of alloys and simple compounds is known. There are numerous articles and a wide selection of literature in which the topic is discussed in detail, and the interested reader should refer to them.

This second chapter only briefly describes the theory of sputter erosion as well as of sputter deposition. The few theoretical views presented here won´t be able to give more than a rough understanding of the topic, but it will be enough to lead from principal considerations to the design features of a sputter deposition setup, that will be adequate to our needs.

2.1 Sputtering

The term 'sputtering' denotes the ejection of near-surface atoms from a material induced by a particle bombardment [beh81]. The bombarding projectiles in this relation can be any sort of particles that can induce an atom ejection, for example molecules, neutral atoms, ions, nuclei, neutrons, protons, electrons and even photons [beh83]. The bombarded material or sputtering target on the other side can be any type of solid.

---

[1] In his patent from 1902 [edi02] he used sputtering to coat phonograph records with a thin layer of gold.

2 THEORY

Sputtering occurs over a wide range of projectile energy, beginning at some electron volts (eV) up to the MeV range and even above. However only sputtering with ions as projectile particles at energies in the eV and low keV energy range is technically easy to realize and relevant for industrial applications. This thesis therefore focuses exclusively on low energy ion sputtering.

Definition Material ejection during the particle bombardment of a targets´ surface is not necessarily the result of sputtering. Processes like evaporation due to beam heating or structural modifications like blistering of flaking also lead to material ejection, but are not denoted as sputtering. Only in the case that even a single ion may in principle lead to an atom ejection, sputtering is given.

Physical and chemical sputtering An atom ejection after ion impact can be the result of a collision cascade induced by the energy and momentum transfer of the ion. It can also be the result of a chemical reaction induced by the ion, that releases energy or generates a chemical instable compound on the surface of the target [beh81]. Both effects are denoted as 'sputtering', the first non-reactive one as 'physical sputtering', the last reactive one as 'chemical sputtering'. In pure physical sputtering, the ejected atoms receive enough energy to overcome the surface binding by a collision or ionisation cascade. In pure chemical sputtering, molecules are formed on the surface of the target due to a chemical reaction between the incident ions and the target atoms, which have a binding energy low enough to desorb at the current target temperature. It is the common principle of physical and chemical sputtering, that enough energy has to be transferred to the ejected atoms, that the binding forces exerted by the target can be overcome. Whether this energy is of kinetic or chemical origin is not of relevance.
Usually pure physical sputtering appears exclusively when noble gas ions are used for bombardment. For reactive ions a mixture of physical and chemical sputtering can be observed, as bombardment of a solid surface with other than noble gas ions usually leads to a chemical reaction between the incident ions and the atoms of the solid. For ion energies in the low eV range this reactions accompany or even dominate the physical sputtering effect. For higher energies the chemical reactions follow the physical sputtering and may even be merely a side effect. To separate physical and chemical sputtering from each other the ejection mechanism is regarded. However the borderline between both is not sharp. For small binding energies, the distinction whether a physical or chemical sputtering effect is present is often difficult [beh83].

2.1.1 Sputter erosion mechanism

Sputtering is initiated by the collision of a bombarding ion with surface atoms of the sputtering target. If the ion is not reflected immediately it will penetrate the target material. Depending on the mass and type of ion and target atoms and depending on the kinetic energy involved, the ion will undergo elastic and inelastic collisions with the target atoms, loose energy and probably be neutralized on its way through the target material. Finally the ion will either come to rest within the target material (i.e. it is 'trapped' respectively it undergoes a chemical reaction) or it will be able to escape through the target surface with reduced kinetic energy. Noble gas ions can generally be trapped only at damage sites in the target material or precipitate as separate phase like gas bubbles [beh83]. Reactive ions can be dissolved or may form a compound phase in the target material or on the surface.

Collision regimes In an elastic collision process the kinetic energy is conserved, which means that the initial kinetic energy of the projectile particle distributes into the kinetic energies of all particles involved in the collision. For metallic sputtering targets elastic collision processes are most important [beh81], as the conduction electrons guarantee a quick distribution of the collision energy between the target atoms. In an inelastic collision reaction kinetic energy is not conserved but to a certain fraction channeled into excitation reactions. For non-conductive sputtering targets inelastic collisions may produce significant numbers of excited electronic states with lifetimes long enough to transfer their energy into atomic motion [beh81] and sputtering reactions.
Regarding the devolution of a collision, it is convenient to distinguish between three qualitatively different situations or 'regimes', that basically depend on the collision energy [beh81] and are shown in figures 2.1 a-f.
For elastic collisions the 'collision regimes' are denoted as 'single-collision regime' (see figure 2.1a), 'linear cascade regime' (figure 2.1b) and 'spike regime' (figure 2.1c), in the case of inelastic collisions the regimes are defined quite similar as 'localised ionisation regime'(figure 2.1d), 'linear ionisation cascade regime'(figure 2.1e) and 'spike ionisation regime'(figure 2.1f).
As we delimit our considerations onto sputtering with ions in the low, medium and high eV range, we have mainly to consider the regimes of single collisions and to some extent linear cascade collisions while sputtering conductors respectively the localised ionisation and to some extent linear ionisation cascade regime when sputtering isolators [beh81]. The spike or spike ionisation regimes could however only occur in the case that we would use very heavy bombarding ions[2].

---

[2] As it will be shown later an Ar plasma will be used, therefore the bombarding ions have an

2 THEORY

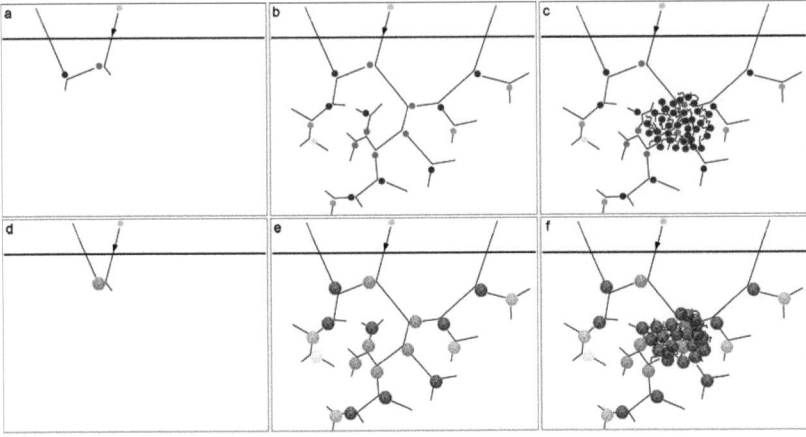

Figure 2.1: Figures by [nor10]: Schematic diagram on the principle evolution of collision reactions in the three collision regimes for conductive (figures a to c) and non-conductive target materials (figures d to f). (a) Single-collision regime: at low kinetic energies the projectile ion penetrates the target surface (horizontal line) and collides with an atom transferring kinetic energy and momentum to it. The trajectory of the ion ends after the collision meaning the ion is stopped now. The primary recoil atom moves through the target and collides with another atom generating a secondary recoil atom and stopping the primary recoil atom itself. The trajectory of the secondary recoil atom points away from the targets surface and as it does not undergo further collisions it is ejected from the target. Collisions of higher orders do not appear. (b) Linear cascade regime: at higher kinetic energies the projectile ion penetrates the target surface and collides with several target atoms generating primary recoil atoms. In the cascade of following collisions secondary, tertiary and higher orders of recoil atoms are generated and several atoms can be ejected. Collisions between the moving atoms are however not frequent. (c) Spike regime: for even higher kinetic energies or large projectile masses the different orders of recoil atoms cannot be determined any more as all atoms in a certain volume are in motion and there is a multitude of collisions between them. The cascade transforms to the so-called spike. (d) Localised ionisation regime: at low kinetic energies the projectile ion penetrates the target surface and deposits a major part of its kinetic energy during an inelastic collision into a small volume that is ionized. The deposited energy in this volume is sufficient to eject an atom from it, but no further collisions or ionisations appear. (e) Linear ionisation cascade regime: at higher kinetic energies the projectile ion penetrates the target surface and undergoes several inelastic collisions, that each produce ionized areas. The energy deposited in this areas is sufficient to eject atoms or to generate areas of secondary or higher order ionisation, that again can eventually eject atoms. (f) Spike ionisation regime: for even higher kinetic energies or large projectile masses the different ionized volumes in the cascade merge and form a large, ionized volume, the ionisation spike.

Sputtering

An algorithm that simulates ion sputtering in the relevant regimes based on a quantum mechanical treatment is implemented in the program SRIM, which was developed by Ziegler and Biersack [zie84], and is described in the appendix.

Atom ejection If an atom receives enough chemical or kinetic energy to overcome the local binding forces it can be considered as energetically unbound. Depending on the position of the atom in the solid and depending on the direction of its momentum vector, the atom usually has to undergo several collisions with surrounding atoms first until it reaches the surface and can possibly leave the solid. Only if an atom reaches the solids´ surface, has a momentum vector that points away from the solid into the vacuum, and is still energetically unbound after all the collisions, it is ejected. Otherwise it is again bound to the solid.

The energy that is necessary to eject an atom from a solid is dependent on the local binding energy as well as on the position of the atom. Generally, atoms on the target surface need less energy to be ejected, as they do not have to undergo collisions to reach the surface and they are weaker bound to the target as bulk atoms. The energy needed to eject surface atoms is therefore an energy threshold, below which no sputtering takes place. If the energy insertion by a collision or chemical reaction is equal or slightly above the threshold, the ejection of surface atoms will begin. For higher energies also an ejection of atoms from the solids bulk is possible.

Usually sputtering energy thresholds are in the range between 20 and 50 eV, depending on the material and depending on the type of bombarding ion. Exact analytical calculation of energy thresholds is unfortunately not possible. They have to be measured experimentally or can also be approximately determined in simulations, for example with the program SRIM (see appendix A1). If only low energetic ion bombardment and the ejection of surface atoms is considered, it is also possible to roughly approximate the energy threshold with the surface binding energies.

2.1.2 Ejected particles

The particles ejected from a target surface during sputtering are mostly neutral atoms with a broad angular and energetic distribution [beh91]. The so-called sputtering yields are used to describe these distributions.

Total sputtering yield The ejection efficiency of a sputtering reaction is measured by the total sputtering yield S, which is defined as the average number of

---

atomic mass of only about 40 AMU. The spike regime will not appear for this mass and energies.

2 THEORY

atoms ejected from the target per incident projectile particle [beh81]:

$$S = \frac{ejected\ atoms}{projectile\ particle} \qquad (2.1)$$

Total sputtering yields may reach values between 0 and 10,000 [joh04], but typically they lie between 1 to 5 [beh81].

The total sputtering yield is dependent on the projectiles´ mass, its kinetic energy and incident angle as well as on the masses and (surface) binding energies of the target atoms involved. For crystalline targets, also the orientation of the crystal axes with respect to the target surface is relevant. It should be noted, that the total sputtering yield always gives the total number of atoms ejected, regardless their energies or directions. If the latter ones are of importance, the differential sputtering yields should be regarded.

Angular distribution The differential sputtering yield $\frac{dS}{d\Omega}$ is a measure for the number of atoms emitted in a certain solid angle. Investigations of $\frac{dS}{d\Omega}$ are one of the most important sources of information on the fundamental mechanisms of sputtering, as they allow to draw conclusions on the momentum distribution in the single sputtering event.

Sigmund developed a theory for cascade sputtering in random media, that reproduces the measured total yields and angular distributions in the single collision and linear cascade regime quite correctly for amorphous and polycrystalline targets[3] [beh91]. The theory predicts a cosine-like angular distribution for normal ion incidence, which is formally given as

$$\frac{dS}{d\Omega} \propto cos(\theta_e)^\beta, 1 \leq \beta \leq 2 \qquad (2.2)$$

where θ_e is the polar angle of the emitted particle and β is a number depending on material and ion energy, which has to be determined experimentally.

Sigmunds theory assumes spatial isotropy of the collision cascades, which is well fulfilled in the case of atom ejection from deeper layers, which appears in the keV energy range. However in case of surface atom ejection or near-surface cascades, as they appear at eV range energies, the spatial isotropy is not given any more [ste01]. For these low energies, the angular distributions often take an undercosine shape with a $\beta < 1$ (shown schematically in figure 2.2).

If the direction of ion incidence deviates from the surface normal at a fixed energy, an inclination in the ejection distribution can be observed (see figure 2.3a).

---
[3]Sigmunds theory also reproduces a major fraction of the atom flux from monocrystalline targets, and can be applied for them too. However it does not describe the characteristic 'Wehner spots' that appear in low-index lattice directions.

Sputtering

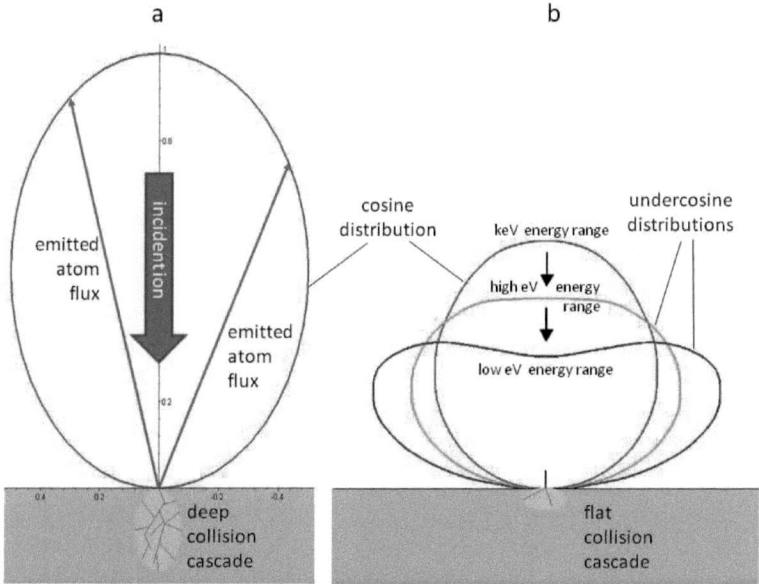

Figure 2.2: (a) Sketch of the angular distribution for the atoms sputtered by normally incident ions. (b) Qualitative change of the angular distribution when the collision cascade is not fully developed. The arrows show how the distribution changes when the ion energy decreases (from [ste01]).

27

2 THEORY

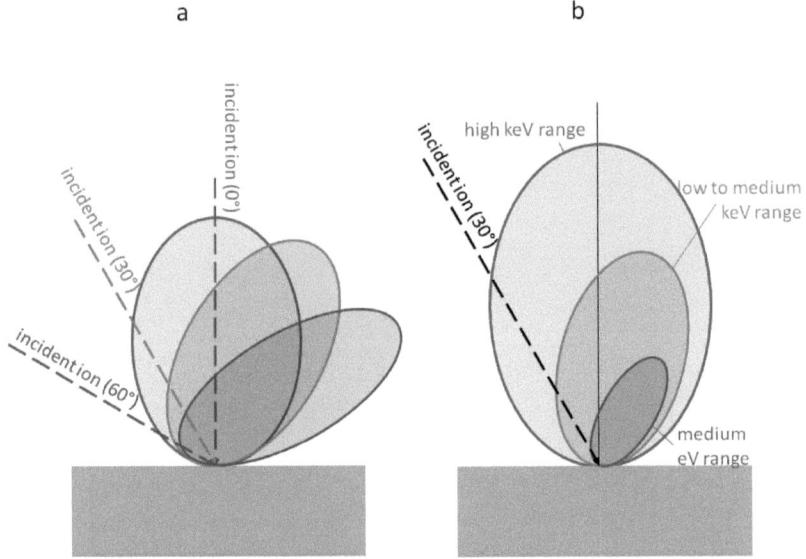

Figure 2.3: (a) Sketch of the angular distribution of atoms sputtered by normal incident ions (0°), as well as ions at 30° and 60° to the surface normal. (b) Sketch of the angular distribution of atoms sputtered by ions at 30° to the surface normal and different ion energy ranges.

The distribution tends to assume an over-cosine shape. For a fixed angle of incidence on the other hand, the observed ejection distributions tend to assume a cosine shape normal to the surface for increasing energy (see figure 2.3b). Both statements are however again only true for an isotropy of the collision cascades. For low ion energies and surface atom ejection, it is hardly possible to determine any general principles in the appearing ejection distributions. They have to be measured in experiment for each combination of ion, target material, incidence energy and angle of incidence.

Energy distribution The differential sputtering yield $\frac{dS}{dE}$ is a measure for the number of atoms emitted with a certain energy. It has been well established experimentally and theoretically, that the energy spectrum of sputtered atoms coming from a fully developed collision cascade is well reproduced by the Thompson

Sputtering

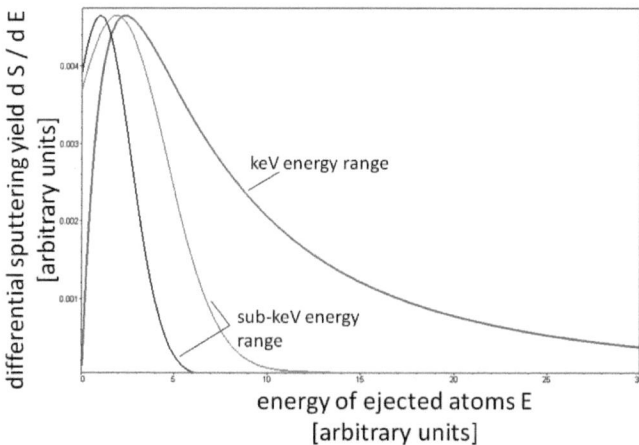

Figure 2.4: Qualitative plot of the energy distribution of ejected atoms for keV ion bombardment (isotropy of collision cascades, Thompson distribution) and sub-keV ion bombardment (non-isotrope collision cascades, sub-Thompsonian distribution).

formula [dep08]:

$$\frac{dS}{dE} \propto \frac{1 - \sqrt{(U_{surface} + E_{atom})/\gamma E_{ion}}}{E_{atom}^2 (1 + U_{surface}/E_{atom})^3}, \gamma = \frac{4 m_{ion} m_{atom}}{(m_{ion} + m_{atom})^2} \quad (2.3)$$

where E_{ion} and m_{ion} are the kinetic energy and mass of the incident ion, E_{atom} and m_{atom} are the energy and mass of the ejected atom and $U_{surface}$ is the surface binding energy of the atom (see figure 2.4).

Experimentally measured energy distributions of sputtered atoms with not fully developed cascade have been observed to deviate from Thompson's prediction, especially for light bombarding ions in the sub-keV range. The peak of the energy spectrum tends to shift to lower energies, the width of the spectrum becomes narrower and the high-energy tail of the energy distributions falls off faster.

Multicomponent yields In case that a sputtering target is composed of more than one species of atoms (also called a 'multicomponent target'), each atom species has its own total sputtering yield as well as its own energy and angular distributions. This multicomponent yields are usually not identical to the sputtering yields of the single components in their pure form.

2 THEORY

2.1.3 Target effects

Continuous atom ejection has a direct influence on the topography of the sputtering target. If the target consists of more than one species of atoms, the sputtering will also change the elemental composition of the target surface and possibly even the elemental composition of the bulk material. The steady ion bombardment will moreover heat up the target. Vice versa will topography, composition and temperature of the target surface have an influence on the sputtering process itself.

Topography change Generally, the sputter ejection of atoms from a surface does not occur uniformly on the ion bombarded area, even if the bombardment is uniformly. This is due to the fact, that the atom ejection during sputtering is a pure statistical effect based on a random collision cascade respectively random ion target reactions. Additionally to that, bombardment and atom ejection continuously modify the surface and the surface near layers. Ions penetrate and leave the target, may react or get implanted, atoms are ejected, displaced and rearranged. Thus, during sputtering the topography of the target surface evolves from its original state toward a new state that is usually very different.
On flat monocrystalline surfaces it can be observed, that small inhomogeneities on atomic scale start to grow and develop into micrometer-sized pyramids, ridges, grooves or holes, depending on the starting conditions, the orientation of crystal and bombardment to each other, the crystal temperature and the type and energy of the bombarding ions. In a flat polycrystalline surface the situation is similar. Each crystallite however behaves as a small monocrystal with unique orientation, and faces therefore unique sputtering conditions and a unique erosion rate, that is different from the surrounding crystallites. The continuous erosion causes by that the crystallites to become visible. Additional grooves or smooth transitions develop at grain boundaries, depending on grain orientation. Thus flat surfaces generally roughen during sputtering.
Very rough surfaces on the other side generally smoothen during sputtering, as prominent domains of the surface will be stronger exposed to erosion while at the same time shadowing less prominent domains. This smoothing continues until the surface roughness is in the order of the grain size (for polycrystalline targets) respectively in the order of the micrometer inhomogeneities (for monocrystalline targets).
The feedback of surface roughness to the sputtering process is a change in the macroscopic atom ejection characteristic of the sputtering target. The macroscopic ejection characteristic is a superposition of the atoms ejected in every single sputtering process on the target surface. The roughness of the surface determines the local angle of ion incidence in every single sputtering process and therefore

Sputtering

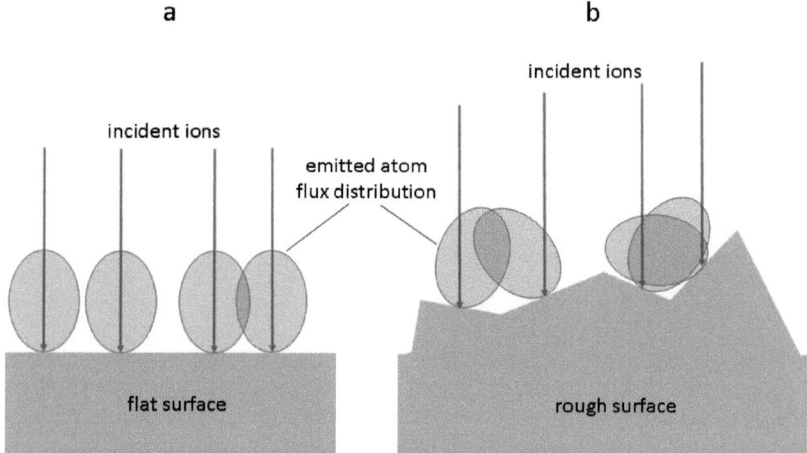

Figure 2.5: Qualitative sketch of single microscopic sputtering reactions and the local angular distributions of the ejected atoms for (a) a flat target and (b) a rough target.

the local angular distribution of the ejected atoms (see figure 2.5). An increase in roughness leads to a randomization of the angles of incidence for ion impact and therefore to a randomization of the different angular distributions of ejected atoms. The resulting macroscopic ejection characteristic is more uniformly distributed and has a less emphasized preferential direction.

Preferential sputtering For a monoatomic target material, the probability for an atom to be ejected during sputtering is determined by several factors as ion type, energy and angle of incidence as well as atom position and local binding forces. The type of target material influences the ejection probability in so far, as it determines the atom positions and the local binding forces. In a material that has more than one species of atoms, the local binding force for an atom of a single species is usually different form the local binding forces of atoms of different species. This means, that even for identical bombardment conditions (ion type, energy, angle of incidence, atom position) the different species will have different probabilities to be ejected during sputtering. This phenomenon is denoted as 'preferential sputtering'.

Preferential sputtering causes the surface composition of a multicomponent target to change, as atom species with a high ejection probability will be depleted on the surface while atom species with a low ejection probability will be enriched. Due to this effect, the total surface fraction covered with low ejection probability atoms will increase while the surface fraction covered with high ejection proba-

2 THEORY

bility atoms will decrease. The surface starting composition changes due to this effect toward an equilibrium composition, which is determined by $p_i A_i = p_j A_j$ $\forall$ i,j, where i and j are different components, p_i is the ejection probability of the i-th component and A_i is the fraction of surface area covered by atoms of the i-th component. After the surface equilibrium is reached, the preferential sputtering effect is exactly compensated by the new surface composition. The different atom species will be ejected according to their fraction in the bulk distribution. Preferential sputtering is therefore just an effect that appears at new targets during the first sputtering, but disappears after the erosion of several atom layers.

Heating and diffusion Ion irradiation heating of the target is a side effect of the ion bombardment during sputtering. Continuous ion bombardment applies a permanent heating to the target, that has therefore to be cooled adequately. Otherwise it will heat up more and more until it starts to melt or to evaporate.

Next to this quite apparent need for cooling, the target temperature also affects the sputtering process. As mentioned in the last paragraph, preferential sputtering changes the surface composition of the target until an equilibrium is reached. If the target is hot enough however, diffusion mechanisms in the target bulk material will appear that avoid this equilibrium to be reached. The depletion of atom species with a high ejection probability at the target surface will cause diffusion processes, that transport further atoms of the high ejection probability species from the target bulk to the surface. This causes a continuous depletion of the whole target.

2.1.4 Sputter deposition mechanism

The atoms ejected from a target during sputtering have kinetic energies in the range of several eV and velocities in the range of kilometers per second. In the usual technical applications, where sputtering is used inside closed vessels in vacuum, these atoms will collide very soon after their ejection with gas atoms or solid structures of the sputtering apparatus like the walls of the vacuum chamber for example. During these collisions[4] the atoms will lose kinetic energy and momentum and finally remain at some surface. If a large number of atoms accumulates on a surface and close to each other they start to nucleate and finally form a solid film.

---

[4]As for the ions in the previous section there is also a certain probability that an atom is reflected from a surface. However even if an atom is reflected several times it will lose its energy sooner or later due to collisions.

Collision If the atoms ejected from the sputtering target hit the surface of a solid substrate, they will be either reflected or bound to the surface. An implantation is unlikely to occur, as the kinetic energy of the atoms is too low. A reflection occurs, if the atom does not loose enough kinetic energy and momentum during the collision with the solids´ lattice; typical duration of this process is in the picosecond range, which is the time of one oscillation of the solids´ lattice. If the atom can however transfer enough of its kinetic energy and momentum into the lattice to come below the surface potential, it will be bound loosely as an adatom to the surface.

At the surface of the substrate a continuous competition between absorption and desorption of atoms takes place. The absorption results from physical surface bonds (physisorption) and chemical surface bonds (chemisorption) and produces adatoms. This adatoms will however desorb again, if they receive enough energy or momentum to overcome the surface potential, perhaps by the thermal vibrations of the solids lattice or by impinging atoms from the gas or from the sputtering target.

Surface diffusion and cluster formation If not desorbed, the adatoms stay on the substrate surface and may have the possibility to diffuse. Dependent on the substrate temperature and dependent on the binding forces an adatom has certain degrees of freedom in diffusion. For strong bonds or a very low substrate temperature no adatom movement will be possible at all. Reduced bond strength or higher temperatures will however allow a diffusion on the surface of the substrate (surface- or 2D-diffusion) or even a diffusion into the lattice of the substrate (bulk- or 3D-diffusion). After some time of migration the adatoms will condensate at some germ.

The reason for adatoms to condensate is an energy gain during nucleation. The single adatoms are only bound to the substrate with the substrates surface potential. By developing further bonds to other adatoms, the total potential for each atom increases and they will become more tightly bound, which lowers their rate of desorption. The condensation of adatoms leads by that to the formation of adatom clusters or so-called nuclei.

Nucleation A cluster can grow both parallel to the substrate by surface diffusion of adatoms as well as perpendicular to it by direct impingement of incident atoms. Depending, whether the adatoms are stronger bound to each other or to the substrate, and depending on the adatom diffusion, the clusters will therefore grow differently. Three different modes of cluster growth are discriminated and shown in figure 2.6.

2 THEORY

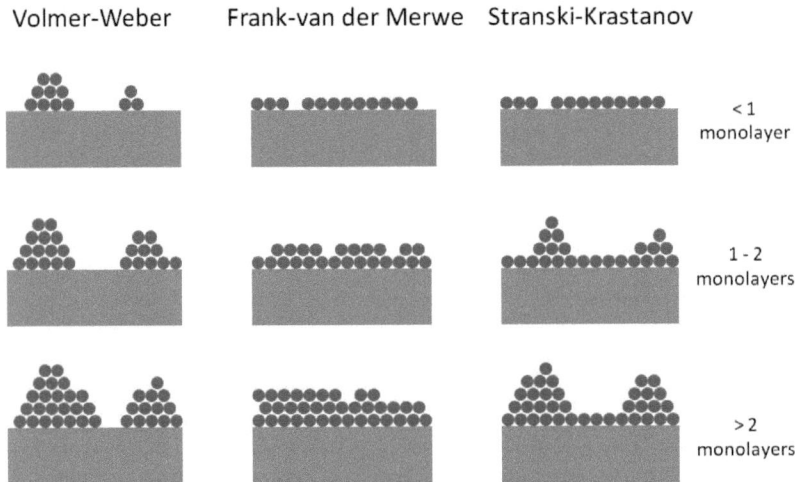

Figure 2.6: Illustration, how the relation of bond strength between adatoms and substrate will influence film growth in its earliest stages. If the bond of the adatoms to each other overweights the bond to the substrate atoms, a nucleus will grow three dimensionally in width an height. This is called Volmer-Weber growth. If the bond of adatoms to the substrate atoms overweights the bond of adatoms to each other, the film will grow two dimensionally as single atomic layers. This is called Frank-van der Merwe growth. If the differences in bond strength between substrate atoms and adatoms is not clearly emphasized, a mixed growth or Stranski-Krastanov growth will appear.

Sputtering

If the bond of nucleus atoms to each other overweights the substrate bond, the nucleus will grow three dimensionally in width and height. This case is called island growth (or: Volmer-Weber growth). It also appears, if the diffusion of adatoms on the substrate surface is slow. In the case the substrate bond overweights the nucleus atom bond or the adatom diffusion is very quick, the nucleus grows two dimensionally as single atomic layers in a layer by layer growth (or: Frank-van der Merwe growth). Usually the differences in bond strength are however not so clearly emphasized and the diffusion speed is neither slow nor fast. Thus nuclei will undergo a mixed growth (or: Stranski-Krastanov growth), that begins as layer by layer growth and starts to form three dimensional islands later on.

A process competing to growth is the breakup of nuclei. It is caused by the nucleus´ surface energy, that increases with the size of the nucleus and causes small nuclei to break apart to reduce both surface and surface energy. Both effects continuously create and destroy nuclei. If a nucleus grows however to or above a certain critical size, the gain in condensation energy will overweight the separating effect of surface energy for every additional adatom. A critical nucleus like that won´t disappear any more but collect further and further adatoms and grow continuously.

The critical nuclei grow in number as well as in size until a saturation nucleation density is reached. The nucleation density and average nucleus size depend on various parameters as energy of the impinging atoms, deposition rate, energies of adsorption and desorption, diffusion rate, temperature, substrate topography and materials involved.

Agglomeration In the next stage of film formation the grown nuclei start coalescing with each other in an attempt to further reduce their surface and surface energy. This tendency to form bigger structures is denoted as agglomeration and the resulting structures are so-called islands. Larger islands grow together, leaving channels and holes of uncovered substrate. The structure of the deposit changes at this stage from discontinuous island type to porous network type. A completely continuous film is formed by filling of the channels and holes.

All stages of microscopic growth finally lead to the formation a macroscopic film on the substrate, if it is continued long enough. The properties of this film are strongly influenced by the grown microscopic structures.

2.1.5 Deposited film

Understanding and control of microscopic growth processes open the possibility of microstructural film engineering, in order to design materials with tailored properties for specific technological applications. Extensive studies of the corre-

2 THEORY

lation between the microstructure of deposited films and the related macroscopic deposition parameters have therefore been carried out in the past decades. The aim of such studies is to get a deeper understanding of how microscopic film growth processes can be influenced by macroscopically controllable parameters. This has led to the development and refinement of structure zone models, which systematically categorize self-organized structural evolution during film deposition.

MD zone model Movchan and Demchishin identified three different microstructures with distinct structural and physical properties, that appear in every film that is deposited in a thermal evaporation process onto a substrate [mov69]. The microstructures are called zones 1 to 3 and result from the microscopic growth processes described in the last section. They were able to show, that these microstructures could be ordered in terms of the homologous temperature T/T_m, where T is the substrate temperature and T_m is the melting point of the deposited material (both in Kelvin).

The low temperature zone 1 structure consists of tapered columnar grains separated by pores or voids. In this zone the adatom mobility is low and incident atoms adhere where they impinge. Microscopic growth is dominated by grain shadowing effects, and voids can occur between the grains due to the missing surface diffusion. At higher temperatures adatom mobility rises and the zone 2 structure with its dense columnar appearance can form. The structure still has a distinct columnar composition, but due to surface diffusion there are no voids between the columns any more. In zone 3, at even higher homologous temperatures, the adatom bulk diffusion starts. Recrystallization and grain formation processes dominate the film growth, and the resulting deposit has an equiaxed grain structure with full bulk material density.

Movchan and Demchishin recognized, that substrate temperature is the dominant factor in the microstructural evolution of films deposited in high vacuum. Based on their observations, they developed their classic structure zone model or Movchan-Demchishin (MD) model (see figure 2.7). For the transitions of zone 1/zone 2 and zone 2/zone 3 in metallic coatings they determined homologous temperatures of 0.3 respectively 0.5, for non-metallic coatings they found similar values. The MD model assumes however, that microstructure evolution and diffusion processes are primarily temperature dependent, which is only true for films deposited by evaporation in a high vacuum. In case of other deposition techniques and a less pure environment, other factors will influence film growth as well and the MD model has to be improved.

Sputtering

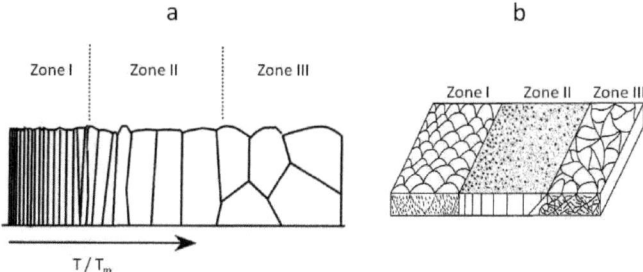

Figure 2.7: Zone model by Movchan and Demchishin (from [mov69]). (a) Cross section of the zone model with the zones 1,2 and 3 ordered according to homologous temperature. (b) Top view of the zone model to emphasize the microstructure.

Thornton zone model During sputter deposition, the substrate and the deposited film are subject to an energetic particle bombardment, that does not appear during evaporation deposition. The bombarding particles are on the one hand target atoms ejected during the sputtering process, on the other hand energetic neutrals, that were generated during ion bombardment by neutralization and reflection of ions from the sputtering target. It was soon recognized, that continuous bombardment of the substrate affects the adatom diffusion processes during film growth, therefore several attempts have been made to include this effects in the MD model.

Thornton investigated the microstructure of sputter deposited metal films and compared them to the structures of the MD model [tho74]. He observed, that for the sputtered coatings the zone 1 structure persisted at higher homologous temperatures with increasing process gas pressure. This is due to gas phase scattering of the bombarding particles, which both reduces the energy of impinging particles and accentuates the atomic shadowing effect during growth. He also observed a fourth microstructure, that consists of densely packed fibrous grains and would be located between the zones 1 and 2 in the MD model. This was denoted 'zone T', as it seemed to be a transition between the two MD zones. Dependent on the gas pressure, the transition between different microstructures could be observed for the homologous temperatures 0.1 - 0.3 (zone 1/zone T), 0.2 - 0.5 (zone T/zone 2) and 0.3 - 0.7 (zone 2/zone 3). These values varied however slightly for all investigated materials.

Based on his observations, Thornton extended the MD model to describe the structure of sputter deposited films. By adding a second axis, he could account the effect of surrounding gas pressure on the film growth (see figure 2.8). The most important point of this extension was to emphasize the pressure as a decisive process parameter, that, together with the substrate temperature, allows to

2 THEORY

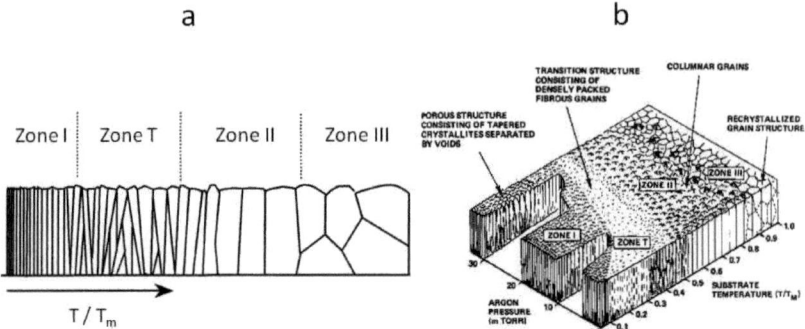

Figure 2.8: Zone model by Thornton (from [tho74]). (a) Cross section of the zone model at one fixed working gas pressure value (here argon) with the zones 1, T, 2 and 3 ordered according to homologous temperature. (b) The top view of the zone model with temperature and the pressure axis maps the observed microstructures.

influence sputter deposition systematically and to grow films with defined properties.

Modern zone models Today several further processing parameters are identified, that affect film growth and morphology. The application of a bias voltage to the substrate generates an additional ion bombardment of the deposited films, that enhances adatom mobility. At high enough bias voltages, this allows to completely suppress the zone 1 structure for all substrate temperatures. By external electric or magnetic fields, the bias ion bombardment can also be directed and adjusted in energy and flux to utilize the atomic shadowing effects to affect texture. Recent experiments also indicate, that even the ion-to-atom ratio incident at the substrate might have an influence on microstructure evolution. Other experiments demonstrated, that impurities in the surrounding gas or in the target material affect grain growth. An increased deposition rate has found to decrease grain size, and columnar grains tend to grow in the direction of the sputtering source.

Every additional processing parameter that is identified to influence film growth can be used to extend the classic MD model and be utilized as a further tool to produce deposits with tailored properties. In this way several different zone models have been developed in the last decades. Other parameters are known to have an influence, but first need to be studied systematically until they can be developed into a zone model. It should be noted however, that the structure zone models are nothing more than a way to order and classify experimental data. A fundamental theory of macroscopic film growth still doesn´t exist.

2.1.6 Substrate effects

Material deposition onto a sputtering substrate usually hardly affects the substrate at all. Vice versa will the sputtering substrate however have a strong influence on the growth of the deposited material and the properties of the emerging film. The zone models have been mentioned already as a tool to classify some of the influences of the substrate on film growth. Others are however not systematically studied yet, but have been mentioned in a rather qualitative way.

Substrate condition The structure zone models described in the last section are intended to predict the influence of processing parameters to the film growth process, especially the influence of substrate temperature. What is not covered by the zone models is however the condition of the substrate prior to the deposition process, as it can hardly be quantified.

The substrate is the basis of the growing film. Its surface therefore strongly affects the growth at least of the first layers of the deposited film. There are three essential factors that have an influence on the film growth process: the atomic structure of the substrate, the topology and the surface contamination.

The atomic structure, that means grain size and crystallographic orientation of the substrate surface, determines the binding energies and nucleation sites of the adsorbed adatoms and by that has an effect on the growth of the growing film [was92]. Generally, a fine grained substrate offers more nucleation sites than a rough grained one, and by that leads to the growth of a more fine grained film with a better adhesion. For a substrate with large grain sizes or even a monocrystalline substrate and in case of crystallographic compatibility between film and substrate, also epitaxial growth may be achieved. In opposite to that, it is possible to grow films with a controlled orientation or even amorphous films on an amorphous substrate.

The surface topology is caused by the grains on one side as well as by surface roughness on the other side. It affects adatom diffusion on the surface, and also influences the deposition and desorption of adatoms by shadowing effects. Generally, a smoother surface allows higher adatom mobility and causes less shadowing, which results in a more uniform growth behavior and denser films with better adhesion.

The microscopic effect of surface contamination is primarily a change in the surface binding, desorption and diffusion energies of the adatoms. These energy changes again influence the film growth process. Depending on the amount of contamination, the growth process may be only slightly disordered on atomic scale or even be completely disturbed up to the macroscopic scale. In the latter case, the resulting films usually show deformed grains, non-uniform growth and a worse substrate adhesion (see paragraph 'Defects' later in this section).

2 THEORY

Atomic structure and surface topology of the substrate can usually be controlled by a proper substrate selection and preparation prior to sputtering. A contamination may result from adsorbed impurities from the working gas, or even from impurities in the sputtering target itself, that get ejected and deposited. A monitoring of the impurity content of the working gas as well as the target material is therefore necessary. Usually, the contamination is however already present on the surface of the substrate before the sputtering process starts, and can therefore also be avoided by a proper substrate preparation.

Reflection The flux composition of atoms sputtered off a multicomponent target does not significantly change between the moment of ejection at the target surface and the moment of incidence at the substrate surface[5]. A composition change appears however, when the impinging atom flux gets deposited on the substrate. As the different arriving atomic species have different reflection probabilities, there will be a selective reflection of impinging atoms from the growing film. The species with higher reflection probability will of course have a lower concentration in the growing film as in the original target material. This composition change can however be calculated accurately and therefore accounted adequately prior to the sputtering process.

Stress Sputter deposited films usually show a certain internal or residual stress, that can be tensile or compressive. If this stress cannot be compensated by the mechanical stability of the substrate, it causes the substrate to bend concavely (for a film with tensile stress) or convexly (for a film with compressive stress). The stress also imposes a shearing force, that may cause the film to break off the substrate, if the film adhesion is too weak to withstand the force.

The residual stress results from two factors [ohr02]: intrinsic stress built up during film growth and extrinsic stress due to thermal effects. The extrinsic or thermal stress is well known from the bimetallic effect. When the film and substrate material are not identical, the film usually shows another thermal expansion behavior as the substrate. Stress appears, when the film is prepared at a temperature different to the temperature where the film is used. It can be avoided however, if the film is already produced at the same temperature it will be used later on.

The reason of intrinsic stress is not fully understood up to now. Several explanations are given in literature, but a universal theory is still not existing. It has been shown, that the incorporation of atoms, for example from the working gas,

---

[5]For distances in the range of several meters or more between target and substrate, a slight change in flux composition might be possible, as different atomic species have different angular ejection characteristics. In typical technical applications, the distance between target and substrate is however small enough that these composition changes can be neglected.

strongly affects intrinsic stress. But also lattice mismatch between film and substrate during growth, variation of interatomic spacing with crystal size, microscopic voids, the arrangement of dislocations, phase transformations and recrystallization processes contribute. Intrinsic stress has shown to manifest itself as tensile as well as compressive stress during sputter deposition. Via the processing parameters it is therefore possible to control the stress.

Defects   An effect, that is closely related to the contaminations already mentioned, are so-called 'defects' of the growing film. Three types of defects are usually classified [bun94]: 'spits', 'flakes' and 'nodules'.
'Spit' denotes liquid target material, that gets ejected from the target surface as small droplets, is thrown upon the growing film and gets incorporated into the film. The bond between the droplet and the surrounding film is usually poor. The droplet may even fall out, leaving a pinhole in the film. The ejection of spits is caused by the eruptive release of gas on the target surface. Gas can be included in porosities in the target bulk material or be generated during the thermal decomposition of impurities.
Spits or other foreign particles that are thrown onto the surface of the growing film lead to the second class of defect, which is denoted as 'flake'. The particle lying on the film faces a higher exposure to the atom flux coming from the target than the general growing film surface. This induces a preferential growth of the film in this area, which is denoted as flake. Usually there is only a marginal bonding between the flake and the surrounding film, so a flake can also fall out, leaving a pit or a crack.
The third type of defect is denoted as 'nodule' and describes cone-shaped structures, that evolve during the film growth process. As spits, they also have a weak adhesion to the surrounding film. They can be caused by asperities on the substrate, but they can also appear without the presence of an asperity [zho98]. The driving force that allows nodule growth is a low adatom surface diffusivity, which is related to substrate temperature. A high degree of scattering of the incoming atom flux due to high working gas pressure as well as a high rate of deposition further increase the nodule size.
The listed defects are usually undesired in the sputtered deposits and should be avoided if possible. Spits can be suppressed, if porosities and impurities in the target material are eliminated, that means if the target material is prepared accurately. Flakes can be avoided, if the presence or impingement of foreign particles on the substrate is prevented. The nodular growth is dependent on substrate temperature and working gas pressure, and can thus be eliminated by choosing appropriate processing parameters.

2 THEORY

2.2 Technology

Up to the present day, the physics of sputtering is not fully understood and still a topic of experimental and theoretical research worldwide. It is well enough understood however to be used successfully in a multitude of technical applications. Generally all of these applications utilize the sputtering effect either to ablate atoms from an objects surface (to clean it or to shape it for example) or to deposit atoms to an objects surface (to coat it or to change its properties) [beh81].

2.2.1 Process

Various methods are described in literature to realize and utilize both physical and chemical ion sputtering for erosion and deposition purposes. It is however not the aim of this thesis, to give a general investigation of all this methods and their applicability in fuel element fabrication. Such a task would go far beyond the feasible scope of a work like that.
We wanted to focus our work on the actual application of functional coatings onto monolithic fuel to demonstrate the feasibility of this method. Therefore we delimited our experimental effort on the realization of a setup, that allows us to utilize the simplest ion sputtering technique necessary for our purpose. We decided to realize an experimental setup, that is able to provide the possibility of physical sputtering as well as reactive physical sputtering. We considered these two processes as sufficient, as they would allow us to deposit all pure elements as well as a variety of simple chemical compounds.

2.2.2 Technical realization

The basic components of a sputtering reactor are denoted as target and substrate. The denotation refers to chapter 2.1, where the target is subject of sputter erosion and the substrate subject of atom deposition. The erosion of the target is started and caused by an ion bombardment, that has to be generated technically. The coating of the substrate is result of the flux of sputtered atoms ejected from the target. Both processes, coating and erosion, thus only need a strong source that provides an intense ion bombardment.
Very effective sources to generate the ion bombardment needed for ion sputtering are gas discharge plasmas. They are technically easy to generate and they constantly produce large numbers of ions, that can be accelerated onto materials by simply applying electric fields. With acceleration voltages of several hundred volts one can induce in this way an intense ion bombardment on surfaces, that leads to a significant sputter erosion and ejection of atoms.

Gas discharge The term gas discharge denotes a process, where an electric current can flow through a gas by the formation of a plasma. The driving force for the current in the discharge is usually an externally applied electric field. Many different types of gas discharges are known, and nearly every type has some technical application. The types can be discriminated by the used electric field (direct field or oscillating field), by the pressure of gas (atmospheric pressure or low pressure), by the temperature distribution in the plasma (thermal or non-thermal) and by their lifetime (self-sustaining and non-selfsustaining) [rai91]. However not every type of discharge is suited to be used as a source for ion bombardment and sputtering in a technical application.

For our particular application we decided to use a glow discharge, which can be classified as a self-sustaining low pressure gas discharge. It can be generated with relatively simple technical means and offers the advantage of continuous ion production together with a homogeneous ion bombardment. We decided against high pressure discharges, as we wanted to avoid the handling of thermal ('hot') plasmas and localized arcs. In opposite to that, the glow discharge produces a non-thermal ('cold') plasma and can be operated both by a direct or an oscillating electric field.

Glow discharge The simplest setup to create a glow discharge is a planar diode, that means an anode and a cathode plate. When a certain voltage is applied to the plates and a gas with an appropriate low pressure is present between the plates, a direct field (DF) glow discharge will evolve. The necessary pressures and voltages to start the discharge are given by the so called 'Paschen curves' for any type of gas and any distance of the electrodes (see figure 2.9). The formation or 'ignition' of a plasma in the diode setup happens within 10^{-7} - 10^{-3}s [rai91], when the 'ignition condition' defined by the Paschen curve is reached. After that the discharge plasma has reached a stable state.

The stable glow discharge shows several regions that are associated with different physical processes (see figure 2.9), which will not be discussed in further detail here. The interested reader should refer to literature for details. For our purpose it is enough to assume the glow discharge plasma as an electric resistance that is located between anode and cathode. An electric current constantly flows from cathode to anode through the plasma volume, and the electric energy is consumed by the plasma resistance.

The charge transport in the plasma works via free electrons and ions, that continuously flow to the anode respectively cathode. The source of the free electrons and ions are ionisation processes in the plasma and at the electrodes driven by the applied electric field. The most important process of this kind is the so called 'vol-

2 THEORY

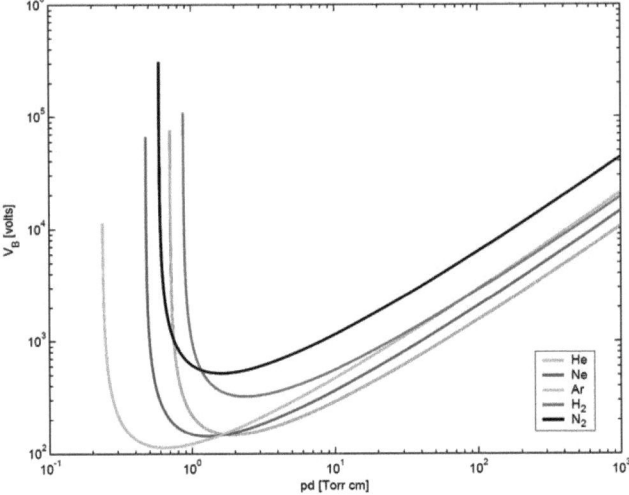

Figure 2.9: The Paschen curve describes the voltage necessary to ignite a discharge plasma in a diode with given distance d between anode and cathode inside a certain gas type with given gas pressure p. The Paschen curve for a particular gas type is dependent on the ionisation energy of the gas and the mean free path of electrons and ions in the gas. The Paschen curves of several common gases are shown here (from [lie05]). For every gas type the curve shows a minimum voltage to ignite a plasma and thus an optimum value of pd. Larger values of pd ('far breakdown') require a larger voltage, as either the electrode distance d rises, which decreases electron acceleration between two collisions, or the pressure p rises, which decreases the mean free path of the electrons and thus effectively also decreases electron acceleration between two collisions. The voltage increase is linearly to pd in this range. Smaller values of pd ('close breakdown') also require an increased voltage, that is however far above linear dependence. This results from the fact, that either the pressure p gets too small, which increases the electron mean free path and thus reduces the collision rate, or the distance d gets too small, which allows the electrons to very quickly reach the anode, which effectively also reduces the collision rate.

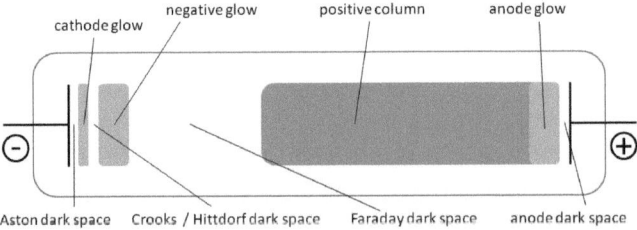

Figure 2.10: Different regions in a glow discharge. Basically, it is distinguished between glow areas, that emit light, and dark spaces, that don´t emit light. Every region is characterized by certain plasma physical effects, that do not appear in the other regions.

ume ionisation', where electrons that have been accelerated in the electric field hit gas atoms and ionize them. The second important process is 'surface ionisation', where accelerated ions hit the cathode and eject electrons. Other processes as the ionisation of atoms by collision with ions or the ejection of ions from the cathode can be neglected, as they have practically no meaning for the glow discharges.

A glow discharge plasma is a continuous and intense source for gas ions, that can be extracted and used for the bombardment of surfaces simply by applying an electrical field to the plasma. In the very simple diode concept it is however not necessary at all to apply an extra electric field to accelerate plasma ions, as an electric field is already applied to generate the glow discharge. The diode field is thus used for two purposes, to generate ions and to accelerate them.

A major drawback of DF glow discharges in a diode is, that both anode and cathode have to be electrically conductive to allow a current in the plasma to flow. Thus if a DF diode is used for sputtering, the target materials are limited to conductive materials only. In case a non-conductive material is used, the surface of the target will accumulate charge, create an electric field of opposing direction to the DF field prevent a further surface ion bombardment. A way to avoid this is a permanent reversion in the polarity of the electrodes. If the frequency of pole changes is chosen high enough, the target surface will not charge up too much and the remaining DF field is still sufficient to allow the glow discharge to ignite. This allows the sputtering of non-conductive materials. The resulting oscillating electric field (OF) glow discharge is very similar to the DF case, as the basic ionisation processes in the plasma stay the same. However there will be no more charge transport through the plasma, and the plasma will not work as a conductor but as an capacitance.

2.2.3 Sputtering setup

According to the assumptions of the previous section, our sputtering setup has to consist basically only of a planar diode, that is a parallel pair of metallic plates. The diode is used as sputtering target and substrate and at the same time as a generator for the glow discharge plasma, that provides the necessary ion bombardment. To reach the low pressure conditions necessary for the discharge, the diode has to be mounted inside a closed pressure chamber, that can be pumped to the required pressures and fed with appropriate working gases. A pumping system as well as devices for vacuum monitoring and gas flow control are mandatory. The voltage necessary to ignite the glow discharge can be generated by using a high voltage power supply.

2 THEORY

Operation The described sputtering diode can be operated both in DF and in OF mode. The operation in DF mode requires less aparative effort but allows only the processing of conductors. As most of the materials relevant to us were metals and alloys, we used DF sputtering for all of our applications. The OF operation is currently investigated and described in chapter 5.

The sputtering process can be operated as a pure physical sputtering process, when a non-reactive working gas is used, or as a physical/chemical sputtering process, when a reactive working gas is used. We wanted to avoid the appearance of chemical reactions during sputtering, thus we used the noble gas Ar as working gas. Reactive sputtering is currently investigated as well and also described in chapter 5.

DF diode If operated in DF mode, the electrodes will be subject of sputter erosion respectively deposition as soon as a glow discharge plasma is ignited inside the sputtering diode. The electric field inside the diode determines the direction of ion and electron movement and thus the direction of bombardment and erosion. As the bombarding ions are positively charged, they will always bombard the cathode. So in a diode glow discharge the cathode is automatically the sputtering target, the substrate is the anode.

Figure 2.11 shows the basic assembly when the diode is operated in DF mode (also called direct current or DC mode). The cathode is continuously eroded and will also face an intense ion irradiation heating. An active cooling of the target will therefore be necessary. The substrate and the inner walls of the pressure chamber will be subject to continuous deposition of material ejected from the target. However only the deposition conditions on the substrate have to be monitored and controlled to achieve desired film growth behavior (especially temperature), the material deposited to the walls is regarded as lost.

Magnetically enhanced glow discharge A DF glow discharge between planar electrodes lead to a uniform ion bombardment of the cathode, that causes continuous and uniform sputter erosion. The deposition rate of usual diode sputtering setups is however rather slow compared to other deposition techniques. An easy way to increase deposition speed significantly is given by the so-called magnetron sputtering process. It uses an additional magnetic field, which is superposed to the electric field in the sputtering diode. The magnetic field forces the electrons in the plasma to gyrate and move according to the magnetic field lines due to Lorentz force[6], and by that confines many electrons in regions of high magnetic flux. The chance for working gas atoms to be ionized in this regions

---

[6]The ions are forced to gyration as well, but their gyration radius is much larger and can therefore be neglected.

Technology

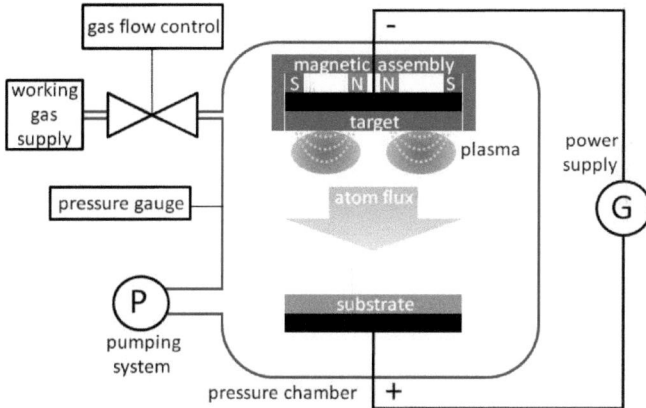

Figure 2.11: Basic assembly of a diode sputtering setup for DF operation. The diode is mounted inside a pressure chamber with pumping system P and devices for vacuum monitoring and working gas supply. The high voltage needed for discharge ignition is provided by a power supply G. On the surface of both anode and cathode the actual target and substrate materials are mounted. They can be exchanged after the process. A magnetic assembly on top of the cathode provides a magnetic field on the target surface. It confines electrons and increases the discharge plasma density, which leads to a significant increase in erosion rate as well as a decrease in working gas pressure.

2 THEORY

increases drastically, which leads to a higher degree of ionisation in the whole plasma and of course to a more intense ion bombardment of the target.

The simplest and most commonly used design for magnetically enhanced glow discharges is the plane magnetron design, where permanent magnets are mounted on the back side of the cathode to get a strong magnetic field on the front plasma facing side. The magnets are usually arranged in a double ring shape, that creates a magnetic field in shape of a torus on the cathode front side. During operation a high density plasma ring forms in the field, and causes massive erosion in the target regions close to it.

The magnetically enhanced glow discharge leads to much higher erosion and deposition rates, operates at much lower working pressures and voltages and allows to define the regions on the target where erosion takes place. It has however the disadvantage, that the target is not eroded homogeneously and the substrate is not coated uniformly any more.

Pressure chamber, vacuum system   Low pressure glow discharges can be operated in the pressure range of 10^{-2} - 10^1 mbar. Above this pressure range, the glow discharge collapses into an arc discharge, below this pressure range the discharge cannot be ignited. If magnetic electron confinement is used, the pressure range can be extended to a lower limit of 10^{-4} mbar. All these processes are however in the low vacuum pressure range, that means they have to be operated in a vacuum vessel.

To achieve the necessary pressures pumps have to be used. For a pressure range of 10^{-2} - 10^1 mbar, the use of a forepump is sufficient. Much lower pressures are required however to reach a pure process atmosphere. Commonly a base vacuum of 10^{-5} - 10^6 mbar is recommended, to achieve a ratio of at least 1000:1 between process pressure and base pressure. This ensures an impurity content of $< 0.1\%$ in the working gas and avoids unwanted chemical reactions in the process. To reach a base vacuum of 10^6 mbar, a turbomolecular or some equivalent pump has to be used however.

Chapter 3

Instrumentation

Based on the considerations of the previous chapter, we built two sputtering reactors, one for basic tests on sputter deposition and erosion and to study processing parameters, the other one to actually demonstrate the feasibility of sputter processing for monolithic U-Mo fuel foils in full size. This chapter describes the construction, the properties and the operation of these reactors.

3.1 Construction

As described in chapter 2.2.3, we decided to build our sputtering reactors as ion sputtering reactors, that utilize a magnetically enhanced glow discharge in a planar diode. First we built a small experimental reactor, that will be denoted as 'tabletop reactor', and installed it into a fume hood inside a radioisotope lab of the TUM Institute for Radiochemistry (see figure 3.3). It was primarily used to study processing parameters. It however turned out very soon, that it can also be used to produce samples for various experimental purposes as well. So we also started to produce tailored samples for different research projects in the reactor as well (see the chapter 4.3).

In parallel we built a large ion sputtering reactor, that was supposed to demonstrate sputter processing for monolithic U-Mo fuel foils in full size. It will be denoted as 'full size reactor'. It is constructed similarly to the tabletop reactor but was installed into a glove box, that allows working under inert gas (see figure 3.1 and 3.2). As the reactor was supposed to process actual U-Mo fuel foils, quite some effort had to be put into safety, security and licensing issues until the reactor could be installed and operated in the radioisotope lab of the LMU[1] Physics Section.

---

[1] Ludwig Maximilians Universität München

3 INSTRUMENTATION

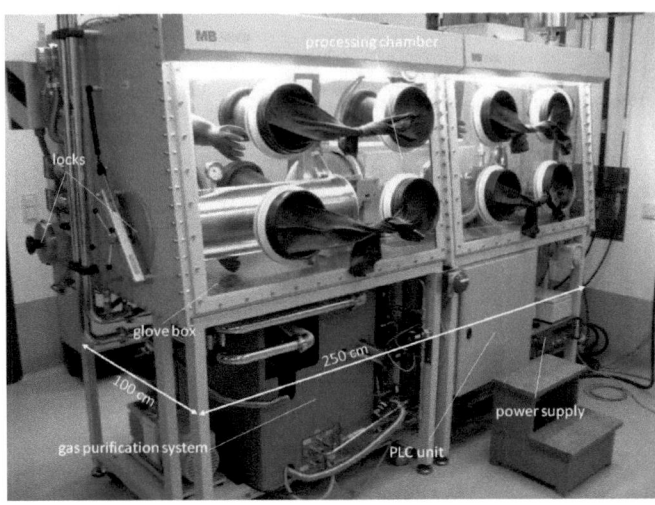

Figure 3.1: Full size reactor inside inert gas glove box in the LMU Physics Section. The whole full size reactor setup contains the sputtering reactor itself, a vacuum system, a gas system, a cooling system, a power supply, an control system, the autarkic gas purification/glove box system as well as several auxiliary systems.

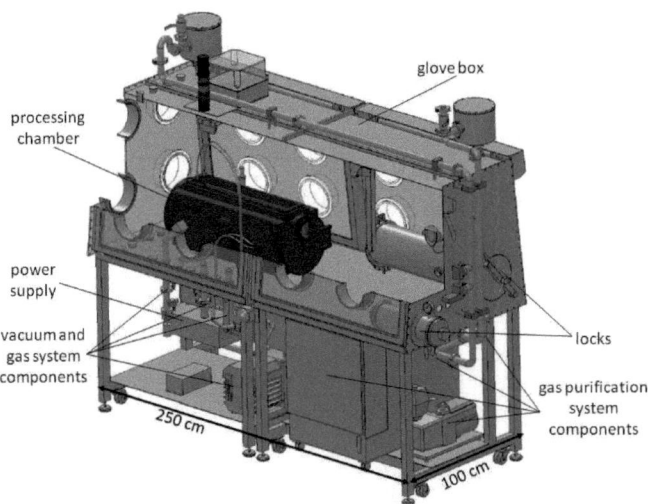

Figure 3.2: Partial cut of the full size setup. The sputtering reactor (black) is mounted inside a glove box. Glove box and all other systems are displayed in gray. The high purity Ar atmosphere in the box avoids oxidation of the substrates or sputtered deposits when the reactor is opened. Moreover it retains contamination and therefore permits to handle radioactive materials.

Construction

Figure 3.3: Tabletop reactor inside a fume hood in the TUM Institute for Radiochemistry. The processing chamber as well as the turbomolecular pump, the vacuum gauge and the needle valve for the adjustment of the working gas pressure were installed inside of the fume hood due to a shortage of space. All other components were installed outside.

3.1.1 Tabletop reactor

The tabletop reactor consists of a sputtering head and a sample table, that together form a diode and are both mounted inside a vacuum chamber (see figure 3.4). The sputtering head consists of a magnetic assembly for electron confinement and structures for cooling and suspension. The sputtering target is mounted underneath the sputtering head. The sample table is composed of a table plate and a cooled table foot. The sputtering substrates are mounted on the surface of the table plate.

Sputtering head The suspension of the sputtering head is a 3 mm thick copper plate (size 160 mm x 80 mm) with a 6 mm diameter copper cooling tube soldered along its edging on top side. The cooling tube forms a frame. In between this frame a magnetic assembly is located.
The magnetic assembly consists of 39 cubic $Nd_2Fe_{14}B$ permanent magnets sized 10 mm x 10 mm x 10 mm with a residual magnetism of $B = 1.43$ T, that adhere onto an iron plate of dimension 148 mm x 68 mm x 10 mm. The magnets are arranged rectangularly in two chains and form a so-called magnetron racetrack (see figure 3.5). The outer chain of magnets consists of magnetic north poles, the

3 INSTRUMENTATION

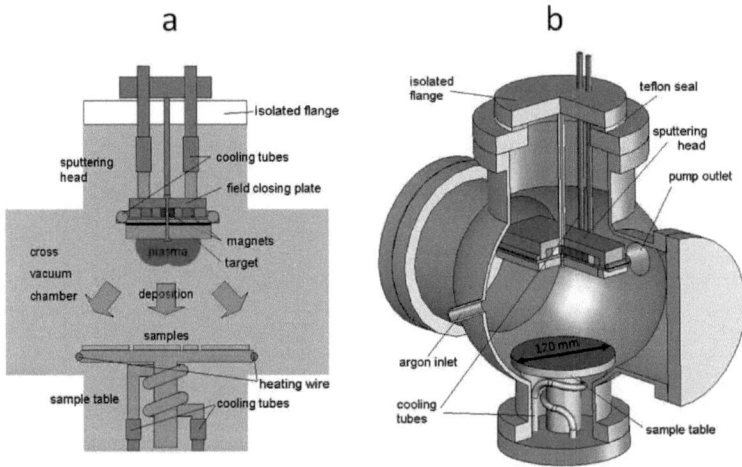

Figure 3.4: (a) Schematic view of the tabletop sputtering reactor. On top the sputtering head with the magnetic assembly (magnets, field closing plate), the cooling tubes and a mounted sputtering target. The sputtering head is at high voltage and isolated from the vacuum chamber. During process a glow discharge plasma erodes the target and deposits the target material into the vacuum chamber. The sample table is equipped with cooling tubes and optionally with a heater and temperature sensors for temperature control. Samples on the sample table can by that be coated at defined conditions. (b) Partial cut of the vacuum chamber with installed sputtering head and sample table that displays gas inlet and pump outlet. The chamber can be accessed only via the front and back flanges.

Construction

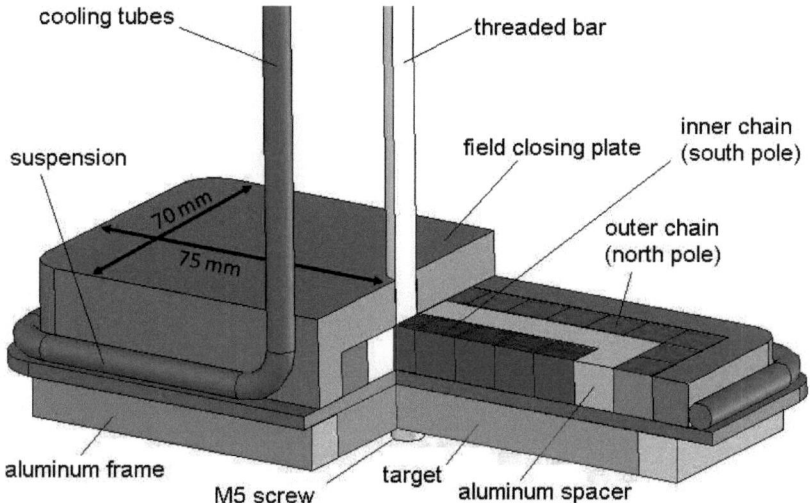

Figure 3.5: Isometric view of the sputtering head with a partial cut to accent the structure. The magnetic assembly formed by 39 $Nd_2Fe_{14}B$ permanent magnets is mounted underneath the field closing plate. The outer chain of magnets forms the magnetic north pole, the inner chain the magnetic south pole of the assembly. Ferromagnetic attraction keeps the magnets in contact with the field closing plate, an aluminum spacer keeps the distance between north pole and south pole magnets. The target is mounted underneath the suspension, that cools the sputtering head. An aluminum frame surrounds the target. It suppresses the side maxima of the magnetic field assembly (see figure 3.6), which is undesired.

inner one of magnetic south poles. Together the assembly of magnets forms a magnetic field in the geometry of an elongated torus, which is necessary for the magnetron enhanced glow discharge described in chapter 2.1.

Figure 3.6 shows a cut of the given assembly with the calculated magnetic field lines. It is clearly visible, that the assembly produces two main maxima and two side maxima. In three dimensions this is equivalent to a magnetic field in the shape of two tori, an inner an an outer one. The smaller, inner torus is formed by the main maxima and located perpendicular beneath the target. This is the desired field structure for the electron confinement that supports glow discharge. The larger, outer torus is formed by the side maxima and located lateral to the target. If not inhibited, this field structure will lead to a second plasma ring, that would sputter the suspension structure. As figure 3.6 shows, it is sufficient to surround the target with an 10 mm x 10 mm diameter aluminum frame to suppress this second plasma ring. The field line distribution of the assembly was calculated with the electromagnetic finite element software VIZIMAG 3.18.

53

3 INSTRUMENTATION

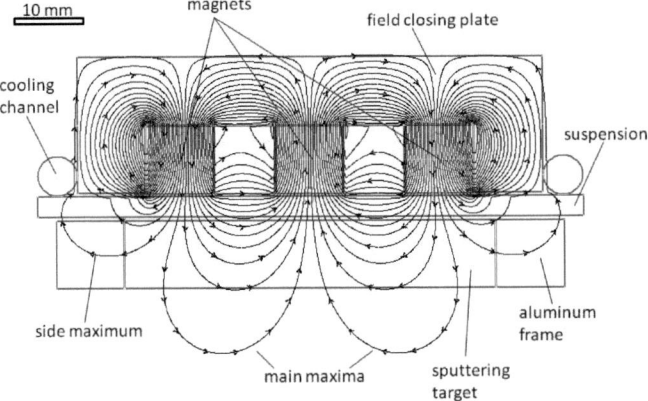

Figure 3.6: Side cut of the sputtering head with field lines that indicate the direction of magnetic force in the field generated by the magnetic assembly (calculated by the program VIZIMAG 3.18). The shielding effect of the field closing plate is clearly visible. The magnetic field lines leave the assembly mainly through the plasma facing side, where the target is mounted. The assembly produces two main and two side maxima respectively a main toroidal field a secondary toroidal field. The main field is located beneath the target. It increases the local electron density by magnetical confinement and by that generates the high density glow discharge plasma needed for the intense ion bombardment of the target surface. The secondary field would generate another plasma ring lateral to the target, that would lead to a bombardment and erosion of the suspension structure. By placing an aluminum frame around the target, the lateral plasma can be suppressed however.

The magnetic assembly is lying upon the cooled copper plate and is surrounded by a copper cooling tube. The good heat conductivity of copper and the proximity of the water cooling assures, that the magnetic assembly always stays below its maximum working temperature[2] of 100 °C.

The copper cooling tube is formed as a twisted loop: it enters the vacuum chamber through the attachment flange of the sputtering head, leading straight down to the copper plate, describes a curve on the copper plate around the magnetic assembly, and leads again straight upwards and through the flange. The sputtering target is mounted underneath the copper plate. It is monolithically shaped with size 122 mm x 54 mm x 10 mm, and affixed by a single M5 screw onto the copper plate above. To guarantee good heat transfer from the target into the copper, a 1 mm graphite foil is placed in between copper plate and target.

By mounting the target underneath the copper plate in the described suspension, one achieves three important things: first, a good cooling of the target by a large-area contact of target and copper suspension and the proximity of cooling water to the target. Second, an adequate magnetic flux through the target without the danger of overheating the magnets. Third, a shielding of the magnetic assembly and the suspension from the erosive effects of the plasma by avoiding a contact between them.

The sputtering head is attached to a DN 100 blind flange via a threaded bar of 150 mm length. The blind flange is electrically isolated to the rest of the vacuum chamber by a special teflon seal, that guarantees good vacuum tightness down to ultra high vacuum range as well as an electrical disruptive strength of several kilovolts. The cooling water flux enters and leaves the vacuum chamber by two liquid media feedthroughs in the flange. Outside the vacuum chamber the cooling water is conducted in isolated flexible tubes.

Sample table The sample table consists of the table foot and the table plate. The table foot is a massive copper cylinder with 40 mm diameter and 80 mm height attached to the vacuum chamber. It is wrapped with several loops of a 6 mm copper tube, that is soldered to the cylinder surface for water cooling. The table foot represents a massive heat sink due to its good thermal conductivity. The table plate is a cylindric copper plate with 120 mm in diameter and 10 mm in height, that is screwed onto the table foot. The samples to be coated are placed on the top side of the sample plate.

Due to the simple construction and mounting of the table plate, it is easily possible to extend or upgrade its functionality. For room temperature coatings a plain table plate can be used. For coatings that require an elevated temperature, an

---

[2]The maximum working temperature of the magnets is given by the manufacturer [che06]

3 INSTRUMENTATION

upgraded table plate with electrical heating and temperature sensor is available. Moreover there are table plates that allow gripping of nine foil substrates sized 25 mm x 25 mm as well as one foil substrate sized 100 mm x 100 mm.

Vacuum chamber Sputtering head and sample table are installed in a vacuum chamber as shown in figure 3.4. The chamber is a stainless steel vessel and provides a DN 160 flange on top and a DN 100 flange on bottom for the sputtering head respectively the sample table. The interior of the chamber can be accessed by two DN 160 flanges on the front and back side of the chamber. During operation these flanges are shut by blind flanges. The process gas Ar is induced via a DN 16 flange on the left side and the vacuum system is attached at a DN 40 flange at the right side of the vacuum chamber. Ar and other gases can be induced into the vacuum chamber by a needle valve, and the process pressure can be adjusted to the needed values.

Vacuum system, cooling water, gas and power supply Figure 3.7 shows the flow scheme of the tabletop reactor setup. The vacuum system consists of a turbomolecular pump in line with a forepump. The forepump is a rotary vane pump, that can provide a base vacuum in the vacuum chamber in the order of 10^{-2} mbar. Together with the turbomolecular pump, a base vacuum in the order of 10^{-6} mbar can be reached. The exhaust gas of the vacuum system is fed into the fume hood, as it might possibly be contaminated.

The high voltage needed for the glow discharge is provided by a DC sputtering power supply with a voltage range of 0 - 1 kV, a current range of 0 - 1.2 A and a power of up to 1 kW. The minus pole of the power supply is connected to the electrically isolated blind flange and the sputtering head (so being the cathode), its plus pole and the ground potential are connected to the sample table and vacuum chamber itself (so being the anode) for a sputter coating of the substrate. For a sputter cleaning of the substrate, the polarity has to be reversed.

The standard working gas Ar is provided by the laboratory Ar line. The working gas pressure inside the vacuum chamber can be controlled by a needle valve and is monitored by a vacuum gauge. In case, that a reactive working gas is needed, it can be fed into the reactor by a second needle valve.

The cooling water for sputtering head and sample table is provided by a temperature controlled laboratory water to air heat exchanger. It is a closed system with a cooling water reservoir of 10 liters.

Construction

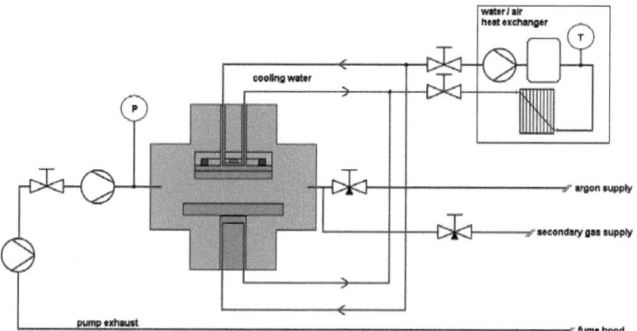

Figure 3.7: Flow scheme of the tabletop reactor setup. The left branch shows the vacuum system consisting of a turbomolecular pump, a forepump, a vacuum gauge and a valve. The right branch shows the working gas supply consisting of an Ar line regulated by a needle valve and secondary gas line regulated by a needle valve. The top and bottom branches are the cooling water flux and reflux tubes of sputtering head respectively sample table. The cooling water was supplied by a laboratory water/air heat exchanger.

3.1.2 Full size reactor

The full size reactor consists of a sputtering electrode and a carrier electrode, that form a diode and are mounted inside a processing chamber (see figure 3.8). The sputtering target is mounted to the sputtering electrode. The electrode contains a magnetic assembly for electron confinement as well as cooling channels. The carrier electrode consists of a water cooled substrate table as well as a cooled carrier frame, that allows to mount sputtering masks and shields.

Sputtering electrode The sputtering electrode is a massive water cooled copper component shaped like a trough (see figure 3.9). It has a size of 792 mm x 154 mm at its fringe and a height of 65 mm. Unlike the sputtering head in the tabletop reactor, which was mounted inside the vacuum chamber, the sputtering electrode in this reactor is itself a part of the processing chamber. It works as a cover of the processing chamber, which has a cavity on its top side that exactly fits with the electrode. The sputtering electrode is lifted into the cavity and finally bears on a special teflon isolating seal. The teflon seal has two important functions: first, it seals the sputtering electrode to the processing chamber so that the chamber is vacuum tight and can be evacuated down to ultra high vacuum range. Second, it isolates the electrode to the grounded vacuum chamber and allows the electrode to be operated at a high voltage of several kilovolts. The concept is basically identical to the teflon seal in the tabletop experiments, but the teflon seal here includes two integrated o-ring seals and is more advanced

3 INSTRUMENTATION

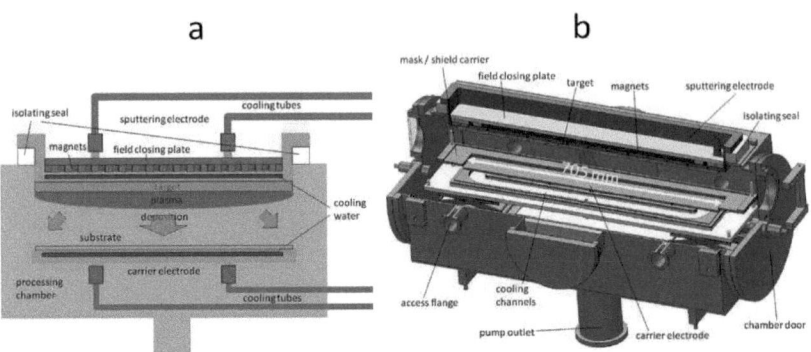

Figure 3.8: (a) Schematic view of the full size reactor. On top is the sputtering electrode, which is operated at a high voltage and therefore electrically isolated from the processing chamber. The magnetic assembly (magnets, field closing plate) is built up identically to the tabletop reactor, but is now located outside of the processing chamber. The water cooling of the sputtering electrode is also done from outside of the processing chamber. The carrier electrode is inside the processing chamber, water cooled and temperature monitored. It provides a carrier frame, that allows to introduce masks and shields into the sputtering process. (b) Partial cut of the full size reactor with installed sputtering and carrier electrode.

from the vacuum engineering point of view.

The sputtering target is mounted underneath the sputtering electrode via 13 M5 screws. The target can be one single monolithic plate with size 702 mm x 122 mm x 10 mm or up to 13 single target elements with size 54 mm x 122 mm x 10 mm each. The size of one of this target elements corresponds to the size of the sputtering targets of the tabletop reactor. The single target elements ('bricks') can therefore be used in the small tabletop as well as in the large full size sputtering reactor.

The bottom side of the sputtering electrode contains wide flat cooling channels that are only about 3 mm away from the top side of the target. They provide cooling to a large area above the target and in parallel prevent the magnetic assembly from getting overheated. To guarantee a good heat contact between target and electrode, a 0.4 mm graphite foil is is placed between the targets and electrode surface. The cooling channels end in two flanged sockets at the atmospheric side of the electrode. From there they are connected via two electrically not conductive flexible tubes to a water feedthrough that leads out of the glove box. The feedthroughs themselves are connected to the central cooling water line.

The magnetic assembly consists of the cubic $Nd_2Fe_{14}B$ permanent magnets sized 10 mm x 10 mm x 10 mm, that were already used in the tabletop reactor. About

Construction

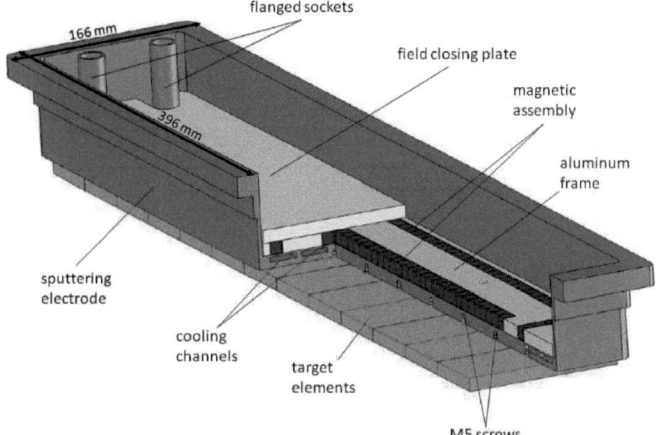

Figure 3.9: Isometric view of the sputtering electrode with a partial cut. The magnetic assembly is formed by about 300 $Nd_2Fe_{14}B$ permanent magnets and shielded by the field closing plate. An aluminum frame keeps all magnets in position. The target consists of up to 13 single target elements and is mounted on the bottom side of the sputtering electrode by M5 screws. Cooling channels inside the bottom side of the electrode cool the target and prevent the magnets from getting overheated. The channels are supplied with water via flexible tubes that are attached to flanged sockets.

3 INSTRUMENTATION

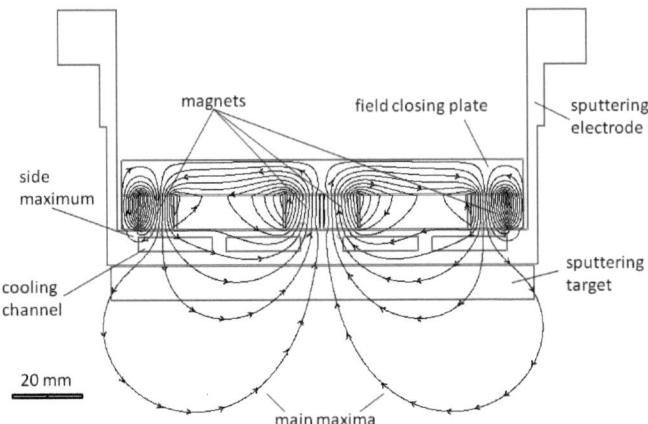

Figure 3.10: Side cut of the sputtering electrode with field lines that indicate the direction of magnetic force in the field generated by the magnetic assembly (calculated by the program VIZIMAG 3.18). Unlike the magnetic assembly used in the tabletop reactor, this assembly produces two strongly pronounced maxima that leave the assembly mainly through the plasma facing side of the electrode where the target is mounted. The side maxima are present, but very much suppressed. The attachment of an aluminum frame or similar constructions is not necessary, as the side maxima always stay within the sputtering electrode.

300 of them are arranged rectangularly in two chains and form a magnetron racetrack (see figure 3.9). A cut view of the toroidal magnetic field geometry is shown by figure 3.10.

Like in the tabletop reactor, the magnetic assembly shows again two main and two side maxima of magnetic flux density. The bigger distance between the inner chain and the outer chain of magnets results however this time in an increased size of the main maxima and vice versa in a decreased size of the side maxima. The change in the magnetic geometry allows therefore to avoid an aluminum frame around the target, as the electrode geometry itself can suppress the generation of a lateral secondary plasma ring.

Carrier electrode The carrier electrode is a water cooled rectangular copper table sized 690 mm x 110 mm inside the processing chamber. It is mounted onto a telescopic tray, that can be completely extended from the processing chamber. Moreover it is adjustable in height, which means, that also the distance between substrate and target can be adjusted. The carrier electrode contains wide flat cooling channels that are only 2 mm away from the bottom side of the substrate. The channels end in two flanged sockets at the bottom side of the electrode. Via two full-metal flexible tubes they are connected to a UHV water feedthrough, that leads out of the processing chamber into the glove box. Another pair of flexible tubes connects the cooling channels to the central cooling water line.

On the carrier electrode a water cooled copper frame is mounted. Its position can be adjusted in height relatively to the electrode, and it serves as a suspension if components like deposition shields[3] or sputtering masks[4] have to be mounted in between the carrier electrode and the sputtering electrode.

Inside the carrier electrode there are five PT-100 temperature sensors mounted 1 mm beneath the surface. They continuously monitor the temperature of a substrate lying on the electrode. For coatings that require an elevated temperature, a secondary substrate plate with electrical heating and temperature sensors can be installed on top of the carrier electrode.

Processing chamber The processing chamber is a 110 liter stainless steel vacuum vessel with a cavity on its top side to mount the sputtering electrode (see figure 3.8(b)). The interior of the chamber can be accessed via two DN 350 flanges, two DN 200 flanges, one DN 100 flange as well as four DN 40 flanges. The two DN 350 flanges, that are at the front and back side adverse of each other, can be sealed with doors. They provide access to the inside of the chamber during substrate mounting. The two DN 200 flanges, located at the left respectively right side, provide additional access in case of repairs. Usually they are however sealed with blind covers. The DN 100 flange at the chambers' bottom as well as the DN 40 flanges on the left and the right side are used for the attachment of vacuum and gas system as well as to feed through cooling water and temperature sensors.

Vacuum and gas system Figure 3.11 shows the flow scheme of the full size reactor setup. Again the vacuum system consists of a turbomolecular pump in line with a forepump, and again a base vacuum in the order of 10^{-6} mbar can be reached. The pumps are connected to the processing chamber via the DN 100 flange in the bottom of the chamber as well as via a DN 40 flange at the chambers side. The chamber is mounted inside of the glovebox, the pumps are mounted outside. Two automatic shutter valves can separate the processing chamber from the vacuum system. The exhaust gas of the vacuum system is filtered twice and fed into the laboratory exhaust line, as it might possibly be contaminated.

The Ar atmosphere inside the glove box is continuously purified and provides an impurity content of ≤ 1 ppm O_2 and H_2O, which corresponds to a gas quality of industrial Ar 6.0. For cost efficiency reasons the standard working gas Ar

---

[3]Deposition shields can prevent undesired material deposition onto the walls of the vacuum chamber.
[4]See chapter 5.1

3 INSTRUMENTATION

Figure 3.11: Flow scheme of the full size reactor installed in the radioisotope laboratory of the 'Sektion Physik' in Garching. The processing chamber containing the planar sputtering diode is mounted inside a glove box filled with Ar. The vacuum system is mounted underneath the glovebox and consists of a turbomolecular pump, a forepump, three pressure sensors and three automatic valves. The working gas for the sputtering process is purified Ar from inside the glovebox, that is regulated via an automatic mass flow controller. The glove box itself is an autonomic system. The two systems are connected to the auxiliary devices: a heat exchanger system, the gas bottle repository and the laboratory air system.

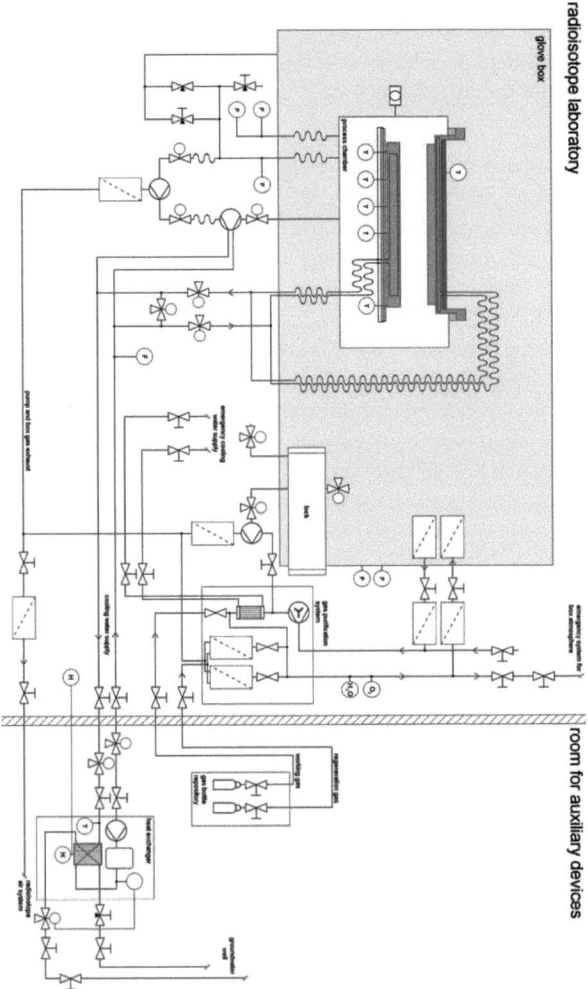

is thus extracted from the Ar atmosphere inside the glove box and not provided separately. The gas pressure in the processing chamber is regulated via an automatic mass flow controller. Reactive working gases can be fed into the reactor externally via a needle valve.

Power supply, cooling water system The high voltage needed to supply the glow discharge is provided by a DC sputtering power supply. It provides a voltage range of 0 - 800 V, a current range of 0 - 30 A and a power of up to 15 kW. The reactor is operated with the sputtering electrode as cathode and the carrier electrode as anode for a sputter deposition on the substrate. For a cleaning of the substrate by sputter erosion, the polarity can be changed by an automatic relay.
The cooling water for the full size reactor is provided by a 20 kW self regulating water to water heat exchanger system. It is a closed system with a cooling water reservoir of 80 liters (see 3.12).

Glove box, gas purification system The glove box is filled with a high purity Ar atmosphere, which is circulated and regenerated by a gas purification system. The oxygen and a water impurity content is continuously kept below a value of ≤ 1 ppm each. The glove box guarantees, that not only the sputtering process itself but also related oxygen sensitive processing steps as the handling of cleaned and coated substrates can be done in an inert atmosphere. A surface pollution due to oxidation can therefore be avoided almost completely. Furthermore, the box allows to handle harmful substances in much larger amounts as it would be possible in a fume hood[5].
The glove box together with the gas purification system form an autonomic installation. They have an automatic pressure and atmosphere regulation, and an own and independent programmable logic controller (PLC) unit that operates the system.

Electronics The full size reactor is a half-automatized system. Preparational works as mounting the sputtering target, preparing the substrate, removing the coated substrates and cleaning of the setup have to be done by operating personal. The processing steps themselves as pumping the processing chamber, adjusting the working gas pressure as well as start, control and monitoring of the sputtering process are done automatically by a PLC unit.
Processing chamber, power supply, vacuum and gas system as well as the cooling

---

[5]During our work we identified three types of substances, that represented a risk during handling: toxic substances (U, Cd), highly or self inflammable substances (U, Zr) and radioactive substances (U).

3 INSTRUMENTATION

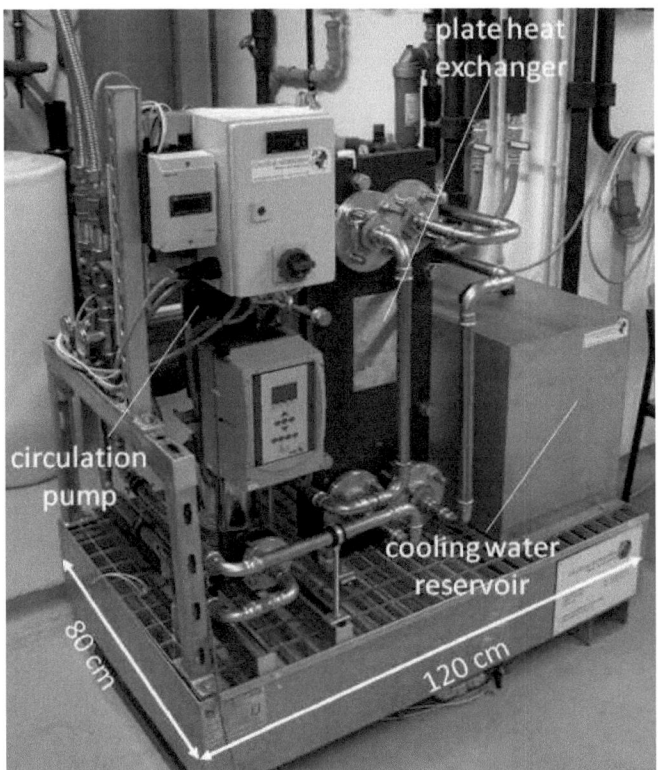

Figure 3.12: Heat exchanger system for the full size reactor setup installed in an auxiliary room of the radioisotope lab. The cooling circuit of the full size reactor setup is a closed system with a 80 liter reservoir of deionized water. A pump drives the water circulation in the circuit. The process heat is removed over a plate type water to water heat exchanger and released into a cooling water stream coming from a groundwater well and going into another groundwater well.

Operation

processing parameters	reactor properties	target properties	substrate properties
gas pressure	plasma geometry	material	material
gas composition	diode geometry	topography	topography
plasma voltage	ion flux	purity	pollution
plasma current	ion energy	homogeneity	microstructure
substrate temperature	ionization efficiency	microstructure	
substrate bias voltage			
target temperature			

Table 3.1: Factors, that have an influence on the deposition process inside a sputtering reactor.

water system with the heat exchanger are controlled and monitored by the PLC unit. Glove box and gas purification system are controlled autonomically.

Auxiliary systems The full size reactor cannot be operated completely autarkic but has to use some systems of the radioisotope laboratory it is installed in. The first of this systems is the radioisotope exhaust-air plant. It holds the pump exhaust of the vacuum system and of the Ar purification system during routine operation. In the case of an emergency it guarantees underpressure inside the glove box to retent harmful materials and radioactivity. Both functions are necessary to assure compliance with radiation protection regulations. The second system is the gas bottle repository. It supplies the full size reactor setup with Ar for the processing chamber and the glove box and ArH for the regeneration of the filters in the gas purification system. The third system is the automated alert system, that is directly linked to the central control room of the TUM campus. As the full size reactor is constructed to process actual nuclear fuel, this alert system is a mandatory requirement to receive a license for operation.

3.2 Operation

A plasma can be ignited between the electrodes of a sputtering reactor, when the correct pressure and voltage parameters are met. The ion bombardment and erosion of the cathode starts immediately, and the sputtered material is dispersed inside the surrounding vacuum chamber.

The sputtering process inside a sputtering reactor is characterized by reactor, target and substrate properties as well as processing parameters (see table 3.1). The properties of deposited film are to a large extent determined by these factors.

3 INSTRUMENTATION

3.2.1 Reactor properties

The ion sputtering process is strongly influenced by the design of the reactor and by the properties of the plasma it can produce. The reactor design includes the diode geometry as well as the geometry of the magnetic assembly. It determines the plasma position, by that the area of sputter erosion and material ejection, the spatial distribution of material deposition and the conditions necessary for plasma ignition and operation. The plasma properties are affected by the degree of electron confinement in the magnetic field and the resulting ionization rate. They determine the plasma density and the rate of sputter erosion and deposition.

Plasma ignition and operation The pressure and voltage conditions that are necessary to ignite a glow discharge plasma in one of our sputtering reactors are given by the Paschen curves (see chapter 2.2). They refer to a glow discharge inside a pair of electrodes with distance d. For small values of d, a large fraction of the material ejected from the sputtering target will reach the substrate surface, as the substrate covers a large solid angle of the ejected material flux. A small value of d is thus desirable. Indeed a too small value of d is fatal for the growth process of the deposited film, as energetic particles from the plasma region will bombard the substrates surface. Indeed we found in experiment, that a value of $d \geq 70 - 80$ mm is necessary to have a largely undisturbed growth process. To be secure, we chose d to be 100 mm for both sputtering reactors. Figure 3.13 shows the measured Paschen curves for different target materials in this configuration. They are valid both for the tabletop as well as for the full size reactor.

Once ignited, the plasma quickly reaches a stable mode of operation with constant plasma voltage and plasma current. The voltage-current (or: U-I) characteristic, i.e. the ratio of plasma current I per applied voltage U, determines the electrical resistance of the plasma. The slope of the U-I curve is a measure for the effectivity of ionisation processes in the plasma and the electron confinement. A small slope means efficient ionization, good electron confinement and the possibility to operate the plasma even at low working pressures. Figures 3.14 and 3.15 show the resulting U-I characteristics for the tabletop respectively the full size reactor recorded for different target materials at different pressures.

Plasma position and erosion area As mentioned in chapter 2.2.3, a magnetically enhanced glow discharge works at much lower pressures than a normal

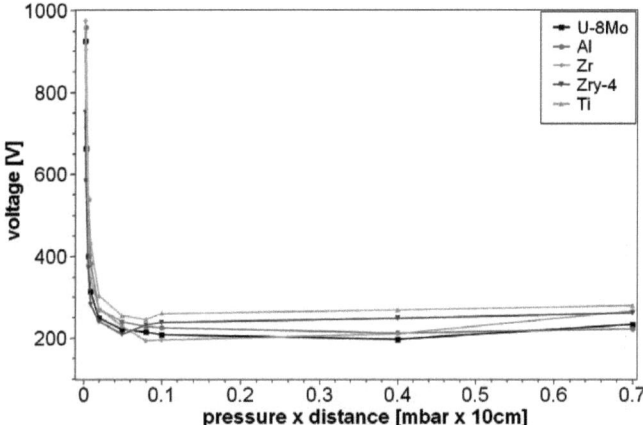

Figure 3.13: Paschen curves recorded for a distance d ≈ 10 cm between the electrodes. This distance is identical for both reactors.

glow discharge. Actually it is usually even operated at pressures, that would not allow to sustain a normal glow discharge at all. This is only possible due to the magnetical field that is applied to the sputtering diode by the magnetic assembly. The position and geometry of this field determines the regions of electron confinement, and by that the regions where a glow discharge can be operated. Outside of this electron confinement region the pressure is too low to sustain a discharge without magnetic field. The magnetical field geometry thus determines, where the glow discharge plasma will be positioned during reactor operation and by that which areas of the target actually face ion bombardment and sputter erosion.

Figure 3.16 left side shows the discharge plasma in the tabletop respectively the full size reactor. The ring shape of the plasmas is clearly visible. The particular magnetic flux densities are shown on the right side of the figure. It should be noted, that the plotted flux densities do not show a ring shape, as they only illustrate the the absolute value of local magnetic field strength, but not the direction of the field. The shape of the discharge plasma is however determined by the trajectories of the electrons gyrating around the magnetic field lines.

The position and density of the plasma determines, which areas of the sputtering target will face ion bombardment and how intense the bombardment will be. As a result, the sputter erosion trenches, that are typical for magnetron sputtering, grow exactly according to the structure of the glow discharge plasma. They can actually be seen as projections of the plasma structure onto the target surface,

3 INSTRUMENTATION

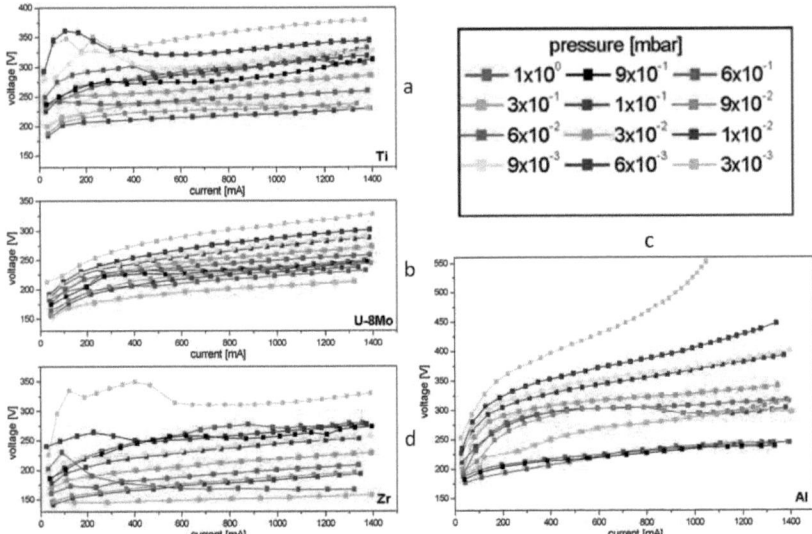

Figure 3.14: U-I characteristics in the tabletop reactor for some available target materials at different pressures. (a) Ti (b) U-8Mo (c) Zr (d) Al. The slope of the U-I curve determines the electrical resistance of the plasma. Figure (b) shows the typical evolution of plasma resistance for a glow discharge: for low currents and voltages the number of free electrons and ions in the plasma is small and the plasma resistance thus large. Higher voltages promote the generation of more free electrons and ions due to ionization and the plasma resistance gets smaller and smaller. The U-I curves in the figures (a), (b) and (d) in general show the same evolution. Unexpected exceptions were however some of the U-I curves in the low pressure range of 10^{-3} mbar and the low current range. The slope of these curves indicates sometimes even negative plasma resistance. The reason for this behavior is not understood yet. Most probably instabilities in the working gas pressure are responsible. It could also be possible however, that the slope illustrates a transition in the gas discharge type, from the instable dark discharge, that can exist for pressures $\leq 10^{-3}$ mbar, to the magnetically enhanced glow discharge, that can exist for pressures $\geq 10^{-3}$ mbar.

Operation

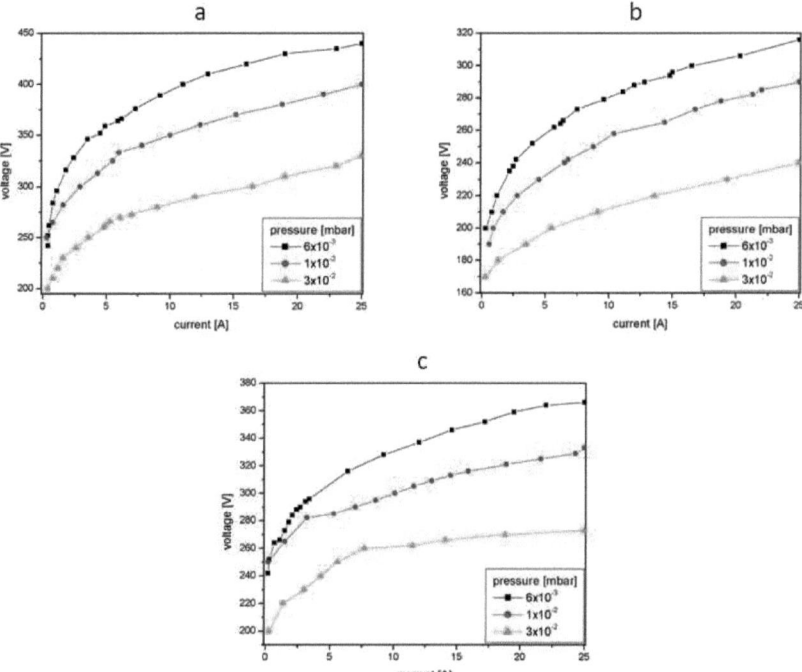

Figure 3.15: U-I characteristics in the full size reactor for some available target materials at different pressures. (a) Al (b) Zry-4 (c) Ti. Unlike the curves measured for the tabletop reactor, all of these curves show the typical evolution of plasma resistance for a glow discharge. This is probably because the working gas pressure inside the full size reactor is adjusted automatically.

3 INSTRUMENTATION

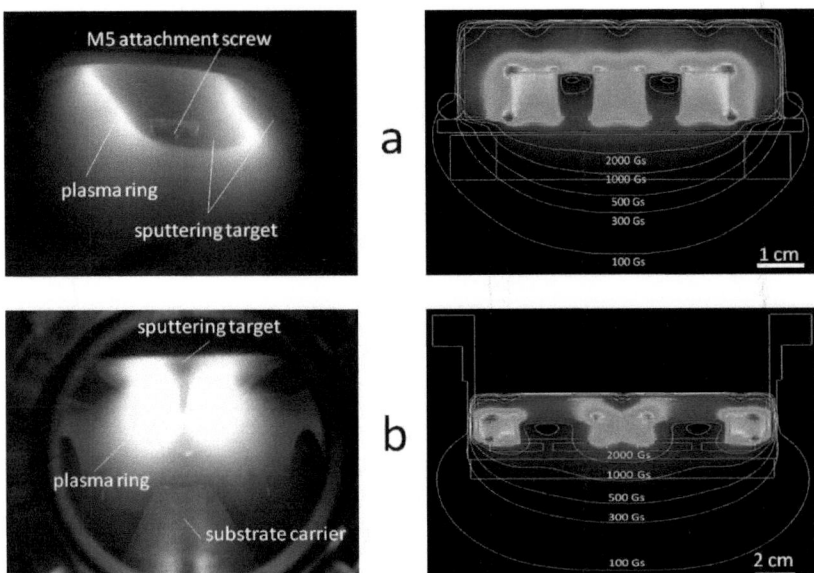

Figure 3.16: (a)Left: Plasma ring during operation of the tabletop reactor. It is generated by the magnetic field of the magnetron racetrack and causes an increased but inhomogeneous erosion of the target. Right: Side cut of the sputtering head with illustration of the absolute value of local magnetic flux density in the field of the magnetic assembly (calculated by the program VIZIMAG 3.18). The brightness indicates the local flux density. The minimum (black) corresponds to values ≤ 1 Gs, the maximum (white) corresponds to 25500 Gs. (b)Left: Plasma ring during operation of the full size reactor. Right: Side cut of the sputtering electrode with illustration of the absolute value of local magnetic flux density in the field of the magnetic assembly. The minimum (black) corresponds to values ≤ 1 Gs, the maximum (white) corresponds to 19000 Gs.

where the boundary of the erosion zone corresponds to the original outer plasma dimensions, and the depth profile of the erosion zone illustrates the density profile of the plasma. The figures 3.17 and 3.18 show the depth profile of the erosion trenches and thus the plasma structure for the tabletop reactor respectively for the full size reactor.

Material deposition The erosion trench in the sputtering target does not only correspond to the position of the discharge plasma, but it also represents the source of material ejection in the sputtering process. The depth profile of the trench accordingly determines the local ejection rate. The material is ejected from the target surface nearly into the complete half space normal to the surface. A fraction of the ejected material directly reaches the substrate and gets deposited there, another fraction reaches the substrate after reflection from the reactors walls. The resulting deposition profile is dependent both from the plasma geometry as well as from the reactor geometry. It is shown in figure 3.19 for the tabletop reactor, in figure 3.20 for the full size reactor, and was gained by thickness measurements of very thick sputtered copper deposits.

The amount of material that gets deposited per time is denoted as deposition rate. It is continuously measured by a rate monitor at one point of the vacuum chamber by an crystal monitor. The measurement principle is the change in oscillation frequency of the crystal that goes along with the mass of material that is deposited on the crystal surface. From the deposition rate at one particular position, that can be measured absolutely, and the material deposition profile, that was measured just relatively and normalized, it is possible to determine the absolute deposition rate at any point of the substrate.

3.2.2 Target properties

The target properties affect the sputtering process itself as well as the properties of the deposit. The target properties relevant for sputtering include target material type, crystallographic structure, purity and homogeneity as well as surface roughness and topography[6].

Binding energy and energy threshold The target material type and crystallographic structure primarily determine the binding energy of the target atoms re-

---

[6]The heat conductivity is of course also an important target property, as the target has to be cooled during operation to avoid melting or evaporation. It has however no direct influence on sputtering.

3 INSTRUMENTATION

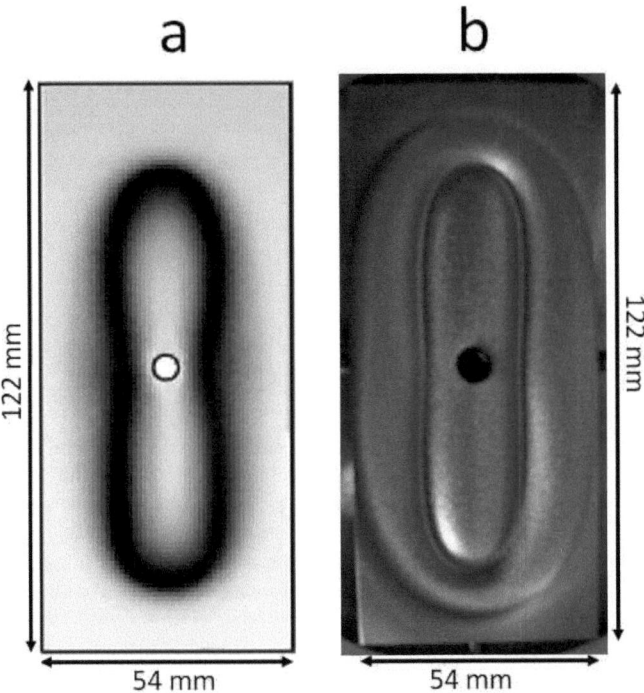

Figure 3.17: (a) The erosion trench in the sputtering target of the tabletop reactor was measured by the optical profile measurement system ATOS at the TUM Institute for Forming and Founding. The brightness indicates the depth of the erosion trench. Black color indicates the maximum depth, while white color is minimum depth. The shades of gray in between correspond linearly to the local depth. The local depth of the erosion trench in the target is also directly proportional to the local intensity of ion bombardment and thus to the density of the glow discharge. The figure thus also shows the actual position and density of the plasma ring. (b) Picture of an actual sputtering target for the tabletop reactor with erosion trench.

Operation

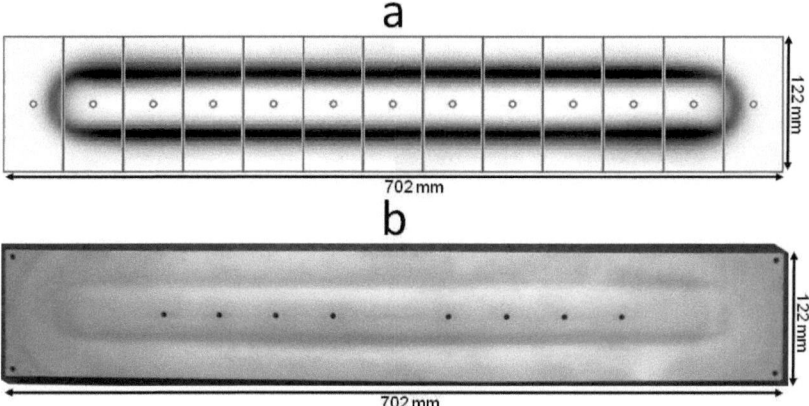

Figure 3.18: (a) The erosion trench in a sputtering target of the full size reactor measured by the ATOS system at the TUM Institute for Forming and Founding. The brightness again indicates the depth of the erosion trench. The arrangement of 13 single target elements to the complete full size sputtering target is illustrated. It is clearly visible, that the plasma ring does not affect the position of the screws that are needed to affix the target elements. (b) Picture of a sputtering target for the full size reactor with erosion trench. The shown target plate does however not consist of 13 single elements but is in this case just one massive plate.

spectively the energy threshold necessary for an atom ejection (see previous chapter). Both parameters were calculated for all target materials that were available to us by the program SRIM (see appendix A1) and are listed in table 3.2.

The energy threshold defines the lower limit of needed ion energy to cause an atom ejection, and by that the necessary acceleration voltage. As it can be seen, it is around 30 eV for nearly all of the materials even though the surface binding energies are all in the range of 5 eV. This is due to the fact, that only a minor part of the energy of the bombarding ions is actually transfered to the ejected atoms, while most of the energy is distributed onto the target bulk.

Impurities and homogeneity As mentioned in 2.1 impurities influence the growth of the deposited film. An absolutely pure target material would be desirable, but usually cannot be achieved at all or only with huge effort and costs. Thus a certain impurity has to be accounted and accepted. The impurity content of the target materials available to us was given by the manufacturers of the single targets and is listed in the last column of table 3.2. In case of multi-component materials - especially alloys - it is sometimes difficult to distinguish between actual impurity content and small amounts of desired additives. What further complicates the situation is the fact, that for many multicomponent

3 INSTRUMENTATION

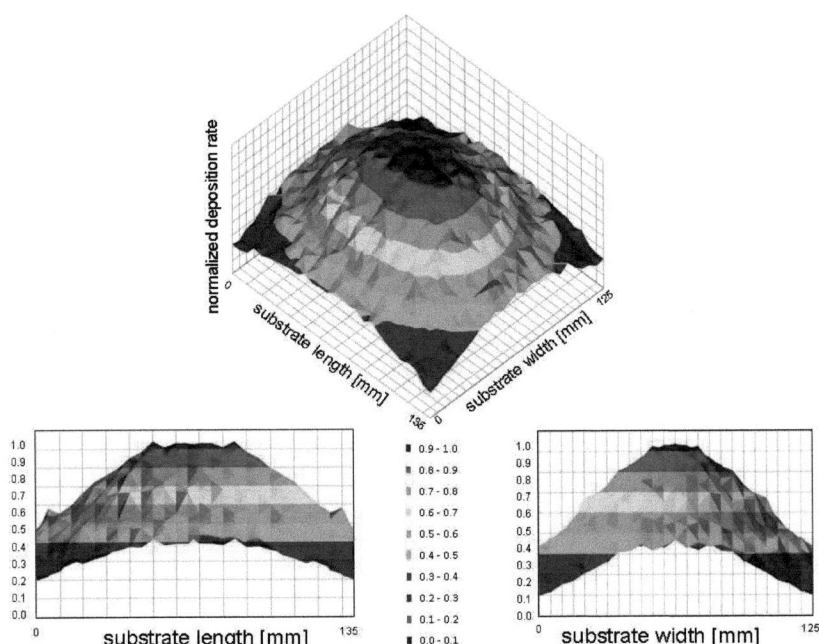

Figure 3.19: Normalized material deposition profile of the tabletop reactor with a substrate target distance of 100 mm. It was gained by thickness measurements of very thick sputtered copper deposits. Top: Isometric view. Bottom: Length and width profile.

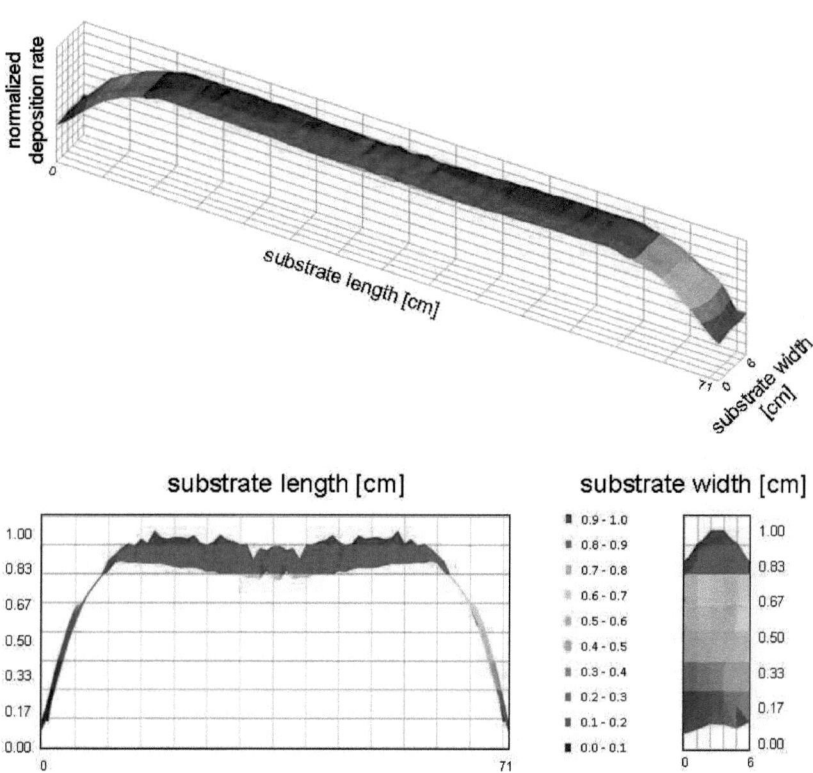

Figure 3.20: Normalized material deposition profile of the full size reactor with a substrate target distance of 100 mm. Top: Isometric view. Bottom: Length and width profile.

3 INSTRUMENTATION

material	surface binding energy [eV]	energy threshold [eV]	impurity [%]
Zr	6.33	30	< 0.07
Ti	4.89	26	< 0.3
Nb	7.59	30	< 0.03
Bi	2.17	47	< 0.01
Si	4.70	20	< 0.0001
Al	3.36	28	< 0.005
Zry-4	6.33 / 3.12	30	< 0.11
U-8Mo	5.42 / 6.83	31	< 0.4
Al-6061	3.36	28	< 0.15
AlFeNi	3.36 / 4.34 / 4.46	27	n.n.

Table 3.2: Surface binding energy and energy threshold for the available target materials calculated by SRIM. In case of multicomponent materials, the surface binding energy is given for the most important constituents, and the energy threshold refers to the first constituent to be ejected. The last column lists the impurity content of the different target materials as given by the manufacturers.

materials there is no uniform composition. Instead, only ranges for the most important constituents are given (see table 3.3).

For multicomponent materials in general, the homogeneity should be considered as well. The term homogeneity refers to the spatial distribution of the constituent elements in the target material. An inhomogeneous distribution of these elements in the sputtering target results in a variation of ejected flux composition during the sputtering process. By that, the single elements will get distributed non-uniformly in the deposited film, which is usually undesired. To get a ho-

material	composition [wt%]
U-8Mo	U (92), Mo (8)
Zry-4	Zr (97.35-98.48), Sn (1.2-1.7), Fe (0.18-0.24), Cr (0.13-0.7), Hf (0.01)
Al-6061	Al (95.85-98.46), Fe (0-0.7), Si (0.4-0.8), Cu (0.15-0.4), Mg (0.8-1.2), Mn (0-0.15), Cr (0.04-0.35), Zn (0-0.25), Ti (0-0.15)
AlFeNi	Al (95.2-97.1), Fe (0.8-1.2), Ni (0.8-1.2), Mg (0.8-1.2), Cr (0.2-0.5), Mn (0.2-0.6), Zr (0.06-0.14)

Table 3.3: Composition of the available multicomponent target materials according to the manufacturers. For most elements only a range of concentration is given. It is thus hardly possible to distinguish between necessary constituents and undesired impurities.

mogeneous elemental distribution in the deposited film, the homogeneity of the target material does however not have to be absolutely perfect. The effect of small inhomogeneities should theoretically become blurred by sputter deposition. Only large inhomogeneities should thus cause a perturbing non-uniform elemental distribution.

Surface roughness and topography The surface roughness and topography of the sputtering target determine, under which angles the bombarding ions will hit the target, how many atoms they will eject, and which angular and energetic distribution the ejected atoms will have. Figure 3.21 shows a fresh sputtering target as we use it in the tabletop sputtering reactor, including macroscopic topography and microscopic roughness. A target for the full size reactor would show a similar surface. As it can be seen, the target is macroscopically plain, but microscopically very rough. The bombarding ions will thus face all kinds of different angles of incidence, and the resulting macroscopic angular ejection characteristic will be a mixture of the various different microscopic ejection characteristics for different angles and different energies.

Surface roughness and topography do not stay constant during the sputtering process, but change with the increasing erosion of the target surface. Figure 3.22 shows a sputtering target with the characteristic erosion trench that forms after several hundred hours of sputter erosion. Again it can be noted, that a target for the full size reactor would show a similar surface structure.

During sputtering the macroscopic topography of the target is continuously changed from a plain surface to a curved surface by the evolution of the erosion trench. The microscopic surface changes from the original roughness to a smooth surface with micrometer-sized conical and pyramidal structures, that are typical for the sputter erosion of polycrystalline materials (see figure 3.23). The changing topography affects the atom ejection characteristics of the target as well. On microscopic level, the bombarding ions still face all different angles of incidence. On macroscopic level, the flux distribution of ejected material will get 'narrower' compared to the flux distribution at beginning of the target life time, meaning that more material is ejected normal to the target surface and less material is ejected at larger angles towards normal. The reason for this shift in ejection distribution are the boundaries of the erosion trench, that have a shadowing effect for material ejected under too large angles.

Redeposition Figure 3.23 also shows a layer of material, that has accumulated on the target surface outside of the erosion area. This material most probably consists of atoms, that get ejected from the target, are reflected from the surface

3 INSTRUMENTATION

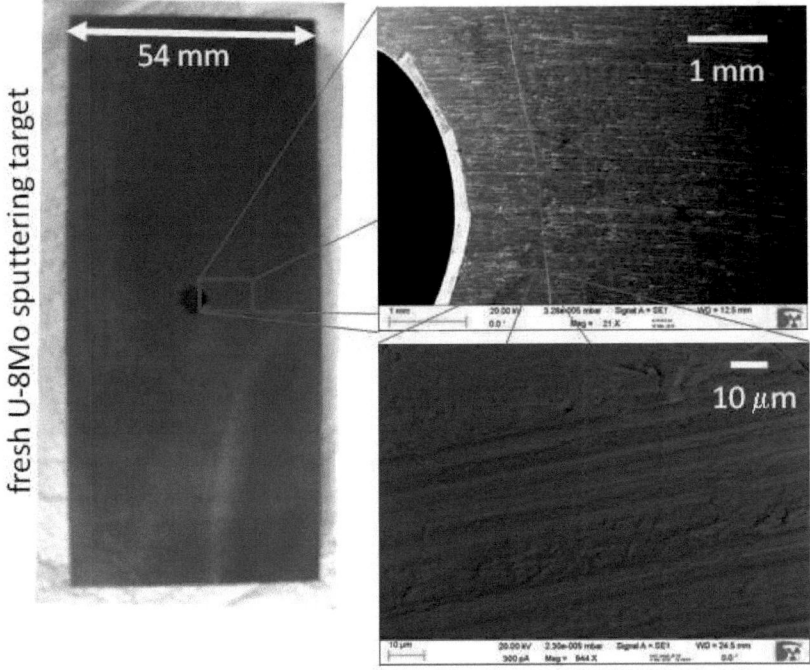

Figure 3.21: Left: Fresh U-8Mo sputtering target for the tabletop reactor. It can be used as a target element for the sputtering target of the full size reactor as well. The surface seems very plain and smooth. Right, top: Target surface with a resolution in millimeter scale. First scratches become visible, but the surface still seems smooth. Right, bottom: Target surface with a resolution in micrometer scale. The surface is not smooth any more but covered with massive scratches that result from fabrication.

Operation

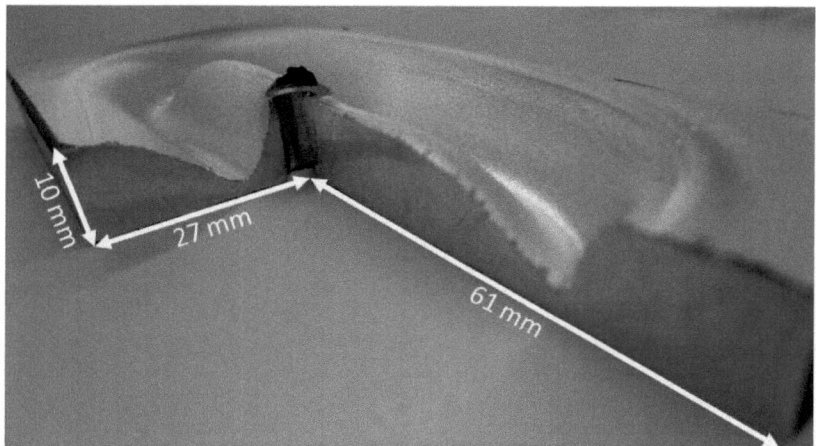

Figure 3.22: Trench profile of the erosion zone in a Cu target. If the erosion zone reaches the bottom side of the target, the sputtering process has to be stopped and the target has to be replaced. Otherwise the suspension structures will be eroded. In our sputtering reactors and with our diode and target geometry, the sputter eroded material amounts about 20 - 25 % of the total target mass for the tabletop reactor and about 30 - 35 % for the full size reactor. The remaining material is not used in the process.

of the substrate respectively deposit, and are finally deposited again on the target surface. We measured the composition of this material and identified it as U-Mo with an significantly increased content of Mo. This is in good accordance to the simulations made in appendix A1, that predict a higher probability of Mo reflection and a lower probability of U reflection from a U-Mo surface and thus an enrichment of Mo in redeposited material.

Cross-magnetron effect A further effect concerning the plasma structure could also be observed on our sputtering targets. In figure 3.24a it can be seen, that the erosion zone at a used sputtering target is not homogeneous but shows areas of increased erosion in the target corners. This effect was first described theoretically by Shidoji [shi00] in 2000, and used by Lopp [lop02] and Fan [fan03] in 2002 and 2003 to explain the anomalous erosion in magnetron arrangements with strong magnetic fields. The so-called cross-corner effect (CCE) or cross-magnetron effect (CME) is basically caused by the used magnetic geometry [dep08]. We use in our reactors a rectangular shaped magnetic racetrack (see figure 3.5 and 3.9), that creates a toroidally shaped magnetic field. For geometrical reasons, this magnetic arrangement yields a distinctly smaller magnetic flux density in the curved sections than in the straight sections. The $\vec{E} \times \vec{B}$ electron current, that always appears

3 INSTRUMENTATION

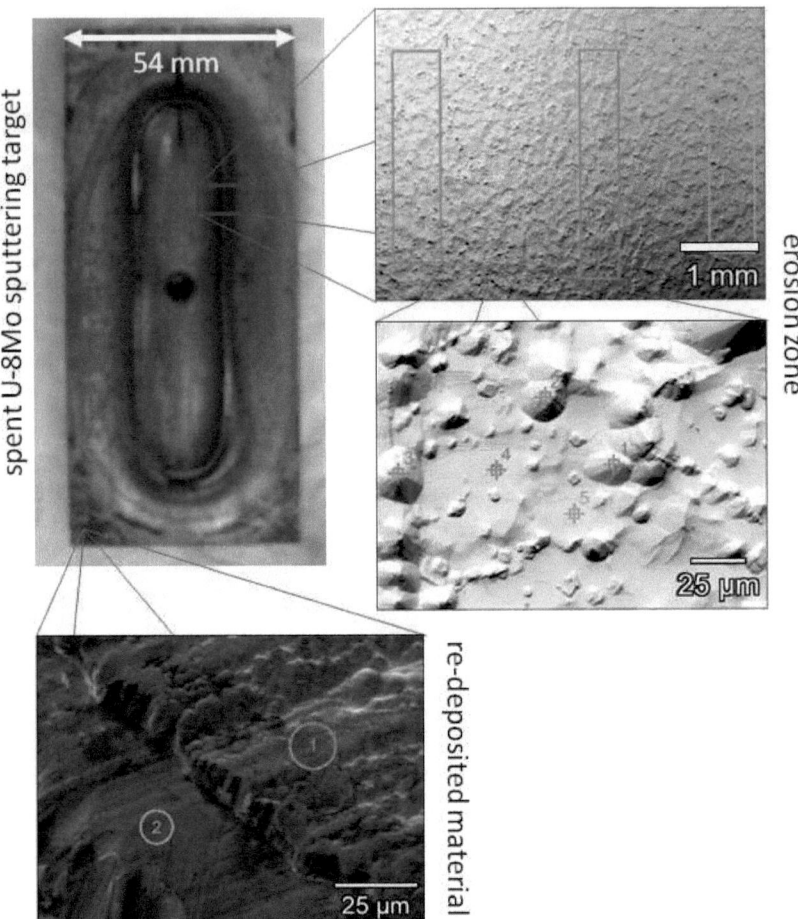

Figure 3.23: Left: U-8Mo sputtering target after several hundred hours of sputter erosion. The surface of the erosion areas has changed from the original very rough and scratchy texture to a rather smooth one (right, top), that shows micrometer-sized conical and pyramidal structures (right, middle). The surface apart from the erosion areas is unaltered, but covered by a layer of apparently re-deposited material (bottom). The target material composition was measured in the labelled regions. For the millimeter scale (right, top) no significant change of composition between target center (1), erosion trench center (2) and target edge (3) was observed. On micrometer scale (right, middle) it was also not possible to resolve any difference in composition between the conical structures (1,2,3) and the plain regions in between (4,5). The layer found on the target edges (bottom) however shows a significant difference between layer material (1) and target bulk material (2), which indicates that the material was apparently redeposited.

Operation

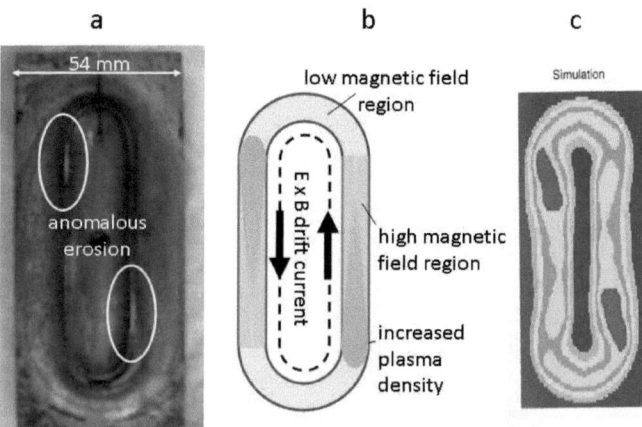

Figure 3.24: (a) U-8Mo sputtering target after erosion. Areas with anomalous erosion behavior are clearly visible (see markings). (b) Scheme to visualize the inhomogeneous magnetic flux density and the $\vec{E} \times \vec{B}$ electron current, that together cause the CCE. (c) Simulation of the anomalous erosion behavior caused by CCE (from [lop02]).

in a magnetically enhanced plasma due to the $\vec{E} \times \vec{B}$ drift, flows along the conductive discharge plasma torus as a closed loop. By an inhomogeneous magnetic flux density the $\vec{E} \times \vec{B}$ current is disturbed, and the electrons accumulate when coming from a low flux region and entering a high flux region. This electron accumulation regions increase the local ionization efficiency drastically and cause the zones of increased erosion.

Next to plasma geometry, CCE imposes a further source of asymmetric deposition behavior in the sputtering reactor. Means to reduce the asymmetry will be shown in section 5.

3.2.3 Substrate properties

The properties of the substrate affect the growth and the properties of a deposited film. The substrate properties relevant for sputtering include substrate material type, crystallographic structure, surface roughness and topography as well as surface pollution. The material type, crystallographic structure and surface pollution of the substrate determine, where and how film growth starts and how good the evolving film will adhere to the substrate. Once the substrate surface is however completely covered with a closed film of several atomic layers thickness, only the microscopic surface roughness and the surface topography will further influence the film growth process.

3 INSTRUMENTATION

Material and crystallographic structure   The kind of application we want to use sputter deposition for does not require a special material or crystallographic structure of the substrate to improve film growth. In fact, the situation is even reversed: a substrate of given material and given crystallographic structure has to be coated without affecting its properties. For us it is thus important, that a substrate will be covered completely, and that the growing films have a good adhesion to their substrate. Thus we only have to consider surface roughness, topography and pollution of the substrates.

Topography and surface roughness   The topography of all substrates we used was macroscopically flat. Curved substrates can in principle be used as well, but shadowing effects resulting from the curvature have to be accounted. For such substrates it might be necessary, to change their orientation towards the ejected material flux during deposition to ensure that the whole surface is covered.

Microscopically rough surfaces will also lead to a shadowing effect, but on microscopic scale. This microscopic shadowing is similar to the grain shadowing effect that is responsible for the formation of the zone 1 structure in the Thornton zone model (see chapter 2.1). An increasing surface roughness will thus increase the total shadowing (consisting of grain and roughness shadowing) and thus promote the formation of film structures according to the Thornton zone 1 [tho77]. The degree of local microscopic surface roughness determines, how large the additional shadowing effect will actually be, and how much it will affect the local film growth at a particular point of the substrate surface.

The Thornton zone model, as it was shown in figure 2.8, does not account for the local influence of surface roughness on film morphology but only gives the morphological development of the growing film in dependency of temperature and pressure. Indeed it is hardly possible to extend the Thornton zone model to a third 'roughness axis', as the term 'surface roughness' summarizes several qualities of the surface topology, and each one is supposed to have an influence on the microscopic shadowing. In general, it is thus only possible to state, that an increasing surface roughness will promote the zone 1 structure, as it is the only one affected by microscopic shadowing.

The microscopic surface of the substrates we used was either polished or grinded, i.e. the surface had a roughness of $\ll 1$ μm for the polished surfaces and < 5 μm for the grinded surfaces (see figure 4.9b). We can thus assume, that the grinded samples could show a tendency towards an increased zone 1 structure, which should be regarded in respect to film growth control (see next paragraph).

Pollution   Surface pollution nearly always worsens film adhesion significantly. The removal of surface pollution turned out to be the major challenge during substrate preparation. The main pollutants we encountered were oxides as well as organic and inorganic contaminations from substrate polishing and handling. The contaminations can relatively easily and sufficiently be removed by chemical cleaning with distilled water, isopropyl alcohol and acetone. The removal of oxides is usually more complicated. Thick oxide layers (≥ 1 μm, see figure 4.9a) were removed mechanically by grinding and polishing. Chemical treatments were investigated as well, but we were neither able to identify appropriate cleaning agents nor a satisfying cleaning procedure. Thin oxide layers ($<< 1$ μm) were only removed if required. We observed, that especially in the production of experimental samples, thin oxide layers can often be accepted and sometimes not even be avoided. In this cases, we did not remove the oxides but coated on the oxidized substrates. If oxidation free surfaces were necessary however, we used sputter erosion before the coating step to remove all remaining pollution (see next chapter).

3.2.4 Processing parameters

The processing parameters are the most important factors during the operation of a sputtering reactor, as their intention is to actively control the sputtering process. Not all of the process parameters listed in table 3.1 are however controllable in our sputtering reactors.

Working gas pressure and composition The working gas pressure in the tabletop reactor can be adjusted via two needle valves. These two valves allow to use and mix two different gases to gain a suited working gas pressure and to control the working gas composition. Each needle valve has an adjustable flow rate of 0 - 5000 $\frac{sccm}{min}$. With the given turbomolecular pump, the working gas pressure can be adjusted between 10^{-4} - 10^{-1} mbar for continuous operation. The standard working gas is Ar 5.0 with an impurity content ≤ 5 ppm. At a working pressure of 10^{-3} mbar, this corresponds to an impurity partial pressure of $\leq 10^{-8}$ mbar. As the base vacuum in the reactor is 10^{-6} mbar, the impurity content of the Ar can be neglected compared to the leak rate of the vacuum chamber. The leak rate itself causes however an impurity content of approx. 1000 ppm during the process. Therefore the working gas composition can be controlled in the order of approx. 0.1 %.

In the full size reactor, the working gas pressure is adjusted by a mass flow controller. The standard working gas is Ar 6.0. A manually controlled needle valve allows to add additional gases to the working gas, before it is fed into the processing chamber. The mass flow controller has an adjustable flow rate of 0

3 INSTRUMENTATION

- 100 $\frac{sccm}{min}$, which allows in combination with the vacuum system to adjust the working gas pressure between 10^{-5} - 10^{-1} mbar. The impurity content of the used Ar is ≤ 0.1 ppm. The base vacuum in the full size reactor is also 10^{-6} mbar, the impurity content due to leaks is however much smaller than in the tabletop reactor, as the whole processing chamber is surrounded by inert atmosphere. The impurity partial pressure during process should therefore be of the order of 10^{-9} mbar or 10^{-4} %.

Plasma voltage and current The plasma voltage and current can be controlled by high voltage supplies. The ranges are 0 - 1 kV and 0 - 1.2 A for the tabletop reactor respectively 0 - 800 V and 0 - 32 A for the full size reactor. The total discharge power is limited to 1 kW respectively 15 kW. Plasma voltage and current can however not be chosen freely, but are dependent on the electrical resistance of the plasma and by that on the working gas pressure. The dependency is described by the U-I-characteristics (see subsection 3.2.1). The plasma voltage on the other hand determines the maximum available acceleration voltage for ions in the glow discharge, and therefore the maximum ion energy during bombardment of the target and the total sputtering yield (see appendix A1). The plasma current is a measure for the number of ions bombarding the target.

Substrate temperature The temperature of the substrate during sputtering process is an equilibrium temperature determined by heat deposition and heat removal. Sources of heat for the substrate are particle bombardment and radiation from the glow discharge plasma as well as eventually integrated electrical heating elements in the substrate holder. Heat is on the other side removed from the substrate primarily by the cooling water. The link between heat sources and heat sink is thus the substrate, or more precisely the heat conduction of the substrate. A good heat conduction will result in good heat removal, and thus in a substrate temperature close to the cooling water level even when a high heat load is applied to the substrate. If the heat conduction is however weak, the substrate temperature might reach values far above the cooling water level even for relatively small heat loads.

The cooling water temperature for the sputtering reactors can hardly be influenced, and the rate of heat removal can thus be seen as constant. The plasma heating can also be seen as constant for a given sputtering process, as plasma voltage and current are kept at constant values. Means for controlling the substrate temperature are thus only heat conduction and artificial heating elements. In case of the tabletop reactor, the heat conduction between the copper substrate table and the water cooled copper table foot can be adjusted by inserting stainless steel spacer disks. There is further a special table plate with an internal heater. Both measures allow to adjust the substrate temperature in the range of 20 - 300 °C. Higher temperatures have not been tested to avoid damages to the setup, but seem possible. If no spacer disks and no heater is used, the substrate temperature will always be around 20 °C during operation.

In the full size reactor it is also possible to insert a steel spacer plate between the water cooled copper carrier electrode and the substrate as a simple mean to worsen the heat conduction. Only by this measure we observed, that substrate temperatures of several hundred °C can be reached. It is also possible to insert a secondary copper substrate plate with electrical heating element on top of the carrier electrode. If no spacer and no heating plate is used, the substrate temperature is usually around 20 °C during operation.

Substrate bias voltage, target temperature Both sputtering reactors do currently not have the possibility to apply a bias voltage to the substrate or to control the target temperature. Both could also be realized if needed. Thus the substrate bias is permanently 0 V in both setups. The target temperature is strongly dependent on the heat contact between target and target cooling and on the plasma power. It covers the range from 100 °C for good contact and/or low power up to values > 1100 °C for bad contact and/or high power.

3.3 Film growth control

As already shown in chapter 2.1, it is possible to control the film growth process to produce a deposit with defined microstructure. But according to Thornton, not only the morphology but also the phases and some mechanical properties of a sputter deposited film are mainly determined by working gas pressure and substrate temperature [tho77]. To reach a certain film morphology, a certain phase or a particular mechanical behavior in a coating of a given material, it is thus necessary to conduct the coating process at a pressure and a substrate temperature that allow the formation of this quality.

Morphology Figure 3.25 shows once again the Thornton zone model for sputter deposited films. As previously shown, working gas pressure and substrate temperature are processing parameters, that can be controlled in our sputtering reactors within certain technical limits. In both of our reactors, we are able to cover the complete pressure axis shown in the figure. We are however not able to

3 INSTRUMENTATION

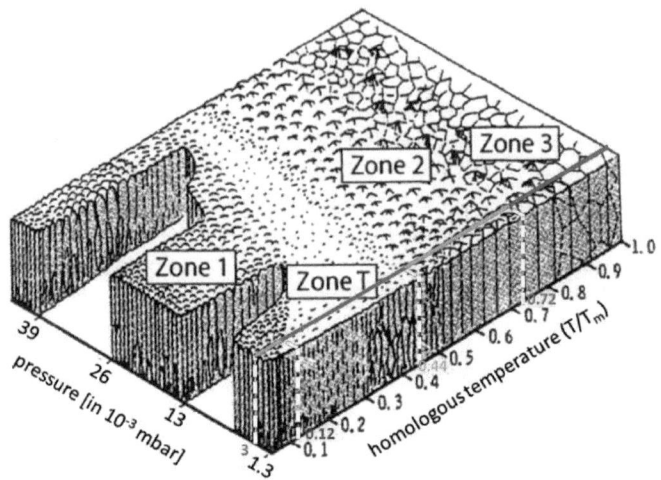

Figure 3.25: General Thornton zone model for the working gas Ar. At a pressure of $3 \cdot 10^{-3}$ mbar the homologous temperatures T/T_m for the zone transitions are given as follows: zone $1/T \approx 0.12$, zone $T/2 \approx 0.44$, zone $2/3 \approx 0.72$. The according temperature values are listed in table 3.4.

cover the complete temperature axis that is shown, especially for materials with a high melting point. Thus not every morphology can currently be reached for every material in our reactors.

Table 3.4 lists the melting points of all the target materials that were available to us as well as the expected zone transition temperatures between the Thornton zone structures 1, T, 2 and 3 at a working gas pressure of $3 \cdot 10^{-3}$ mbar. We usually conducted the sputter deposition at this pressure, as it seems us to be a good compromise between the pressure requirement of the glow discharge (stable and confined in a range between $1 \cdot 10^{-3}$ - $5 \cdot 10^{-2}$ mbar), the attempt to reduce ejected atom scattering in the gas phase (at lower pressures more material flux reaches the substrate) and the aim to avoid the zone 1 structure for a deposition at room temperature.

In the last column of table 3.4 it is listed, which zonal structure can be expected for the deposition of a specific material onto a substrate at room temperature at this specific pressure. Except for Nb, the zone T structure[7] will be reached for all of the listed materials already at room temperature, while the zone 2 or

---

[7] Zone T is characterized as a compact structure consisting of densely packed fibrous grains.

Film growth control

material	T_m [K]	zone 1/T [°C]	zone T/2 [°C]	zone 2/3 [°C]	expected structure @ room temperature
Zr	2130	-17	664	1261	T
Ti	1941	-40	581	1125	T
Nb	2750	57	937	1707	1
Bi	1837	-53	535	1050	T
Si	1683	-71	468	939	T
Al	933	-161	138	399	T
Zry-4	2123	-18	661	1256	T
U-8Mo	1408	-104	347	741	T
Al-6061	925	-162	134	393	T
AlFeNi	922	-162	133	391	T

Table 3.4: Melting temperatures and expected zone transition temperatures for all available coating materials at a working gas pressure of $3 \cdot 10^{-3}$ mbar Ar according to the Thornton zone model. For a deposition at room temperature one expects the zone T structure for nearly all listed materials.

3 structures[8] would require a heating of the substrate. If a zone 1 structure[9] is desired, the substrate has either to be cooled or the deposition pressure to be risen. Furthermore a possible effect of the substrate roughness has to be accounted, that could possibly rise the homologous temperature of the zone 1/T transition.

The decision, which of the four different film morphologies is the best for a particular application, depends on the required density of the film, but also on required phases and mechanical properties.

Phases The substrate temperature does not only affect the morphology but also the crystallographic phase of a sputter deposited film. According to Thornton, the structural order in a coating is produced largely by the mobility of the deposited adatoms [tho77]. Figure 3.26 illustrates this.

Highly disordered, amorphous-like structures are expected, if adatom mobility is negligible and the adatoms come to rest at the point of their impingement. This is the case for very low homologous temperatures T/T_m according to the zone 1 structures. Higher values of T/T_m, that allow a certain adatom surface mobility, promote the formation of equilibrium high temperature phases, which corresponds to the zone T morphology. At more elevated substrate temperatures adatom mobility will finally reach a level adequate to form the phases predicted

---

[8]Zone 2 and 3 are characterized as a dense columnar respectively an equiaxed grain structure.
[9]Zone 1 consists of tapered columnar grains separated by pores or voids.

3 INSTRUMENTATION

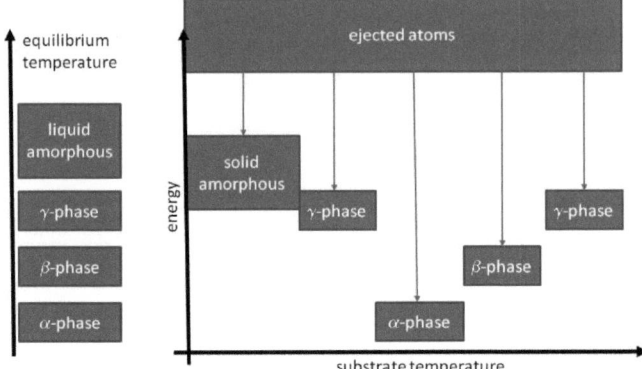

Figure 3.26: Schematic illustration of the influence of the substrate temperature on the crystallographic phase formation in a sputter deposited film (according to [tho77]).

by the equilibrium phase diagram. This is the case for the zone 2 and 3 structures. Table 4.2 lists the expected phases for a deposition at room temperature as well as the phases that would be present if the materials were in thermodynamic equilibrium at room temperature.

It should be noted, that a high temperature phase that is formed on a high temperature substrate during sputter deposition may be lost by bulk diffusion during a too slow cool down. A high temperature phase formed on a low temperature substrate on the other side is relatively stable towards diffusion due to low bulk diffusion rates.

Mechanical properties The morphology determines the hardness, strength and ductility of a sputter deposited film. The zones 1 and T are characterized by a high dislocation density, which is a result of the low adatom mobility during formation. Accordingly, films in these structures tend to have a high hardness and strength, but only small ductility. With increasing substrate temperature the dislocation density decreases. Correspondingly this also decreases hardness and strength, but increases the ductility of the film. The according morphologies are the high zone T and the low zone 2 structure. Films in the high zone 2 or the zone 3 structure finally show the hardness, strength and ductility values of fully annealed bulk materials.

material	phase in thermodynamic equilibrium @ room temperature	expected phase for deposition @ room temperature
Zr	hcp	fcc
Ti	hcp	fcc
Nb	bcc	unclear
Bi	rhomboedric	unclear
Si	diamond cubic	diamond cubic
Al	fcc	fcc
Zry-4	hcp	fcc
U-8Mo	orthorhombic	bcc
Al-6061	fcc	fcc
AlFeNi	fcc	fcc

Table 3.5: Comparison of the crystallographic phases of target materials at thermal equilibrium at room temperature and of the expected phases in films that have been sputter deposited at room temperature. For allotropic materials that get deposited in zone T structure (like Zr, Ti, Zry-4 and U-8Mo) one can expect the formation of the high temperature phases. For non-allotropic materials in zone T or zone 2 structure (like Si, Al, Al-6061 and AlFeNi) the equilibrium phase will most probably be formed. For Nb and Bi it is not clear, which phase can be expected.

Chapter 4

Application

The aim of this thesis was the realization of a setup, that allows processing of monolithic U-Mo nuclear fuel by ion sputtering, and an investigation of potential applications of this technique. For our experiments concerning this matter we built two ion sputtering reactors, that allowed us to investigate the deposition, the erosion and also the coating process of U-Mo. As a primary field of application for this three processes we examined the fabrication of fuel elements for nuclear reactors. During our work we recognized however, that the sputtering reactors are valuable tools for our U-Mo metallurgy projects, as they allow a quick and simple fabrication of small tailored samples for different experimental purposes. The fabrication of scientific samples thus turned out to be a second promising field of application.

4.1 U-Mo processing

Sputtering techniques do not allow to shape a given material to any desired geometry. Basically, they only allow to remove material from surfaces or to deposit material onto surfaces. Therefore any application of sputtering in U-Mo processing is limited to the deposition or removal of material[1]. The following applications of ion sputtering seem feasible and reasonable to us in this context:

- the **formation of a U-Mo film** by sputter deposition
- the **surface cleaning of U-Mo items** by sputter erosion
- the **coating of U-Mo items** by sputter deposition

We tried to test the feasibility of these ideas in our experiments.

---

[1] It would also be possible to modify given surfaces via implantation of atoms. This is a closely related but separate technology and will not be further regarded here.

4 APPLICATION

4.1.1 Film formation

We investigated the sputter deposition of U-Mo in detail at the example of U-8Mo (see also [can11]). We analyzed the composition of a sputtering target of U-8Mo before and after the sputtering process as well as the composition of a deposit produced during the sputtering process.

U-8Mo target erosion Figures 3.21 and 3.23 show a U-8Mo sputtering target before and after the erosion process in the tabletop reactor. The target has been operated at a plasma voltage of 240 V and a plasma current of 1 A. The maximum ion energy was thus limited to 240 eV for single charged ions. The corresponding total sputtering yield for this energy was calculated to be 0.57 atoms/ion for U respectively 0.13 atoms/ion for Mo (see appendix A1).

As explained in chapter 2.1, the preferential sputtering effect will very quickly create a surface layer of altered composition, that grows until an equilibrium is reached. After that the target material will be ejected in its bulk composition. The composition of the altered layer can be calculated from the assumption of equilibrium: U atoms are ejected with a sputtering yield of S_U, Mo atoms with S_{Mo}. Once the surface equilibrium is reached, the U atoms cover a fraction A_U of the surface, Mo atoms cover a fraction A_{Mo}. The surface equilibrium can be expressed in these terms as $S_U \cdot A_U = S_{Mo} \cdot A_{Mo}$. From the simulated sputtering yields (appendix A1) and the atomic radii of U and Mo we calculated, that the altered layer should have a composition of 84.7 at% Mo and 15.3 at% U [2]. As the bulk composition of the U-8Mo is around 20 at% Mo and 80 at% U, the altered layer will face an enormous change in composition.

In EDX measurements of the erosion zone (see figure 4.1), we found a composition of 14 - 15 at% Mo and 85 - 86 at% U, which is close to the bulk composition of 20 at% Mo and 80 at% U but completely different to the composition of the altered layer. The EDX method however measures the average composition in a volume element of 2 μm depth and 1 μm diameter, which covers the altered layer as well as a large fraction of bulk material. Our measurements thus confirm, that the altered layer created by preferential sputtering has to be much thinner than 2 μm, as otherwise the measured composition would be closer to the calculated values. It further confirms, that at our target working temperatures there is no target depletion effect of high yield materials due to diffusion. The deviation of

---

[2]This is value is only true for a monoenergetic ion bombardment with 240 eV ion energy. The ion bombardment in the sputtering reactor has however an energetic distribution with 240 eV as upper limit and 0 eV as lower limit. Thus the formula $\int S_U(E) \cdot A_U dE = \int S_{Mo}(E) \cdot A_{Mo} dE$ has to be solved to determine the real altered layer composition. For energies above 50 eV the ratio S_{Mo}/S_U is however nearly constant (see figure 6.1), and the given value should hence be a good approximation.

measured target composition and expected target composition has to result from other effects however.

We suspected inhomogeneities in the spatial distributions of U and Mo in the target bulk material to be responsible for the measured deviation in composition. To verify this, we cut samples from a U-8Mo sputtering target before and after sputtering. These samples were measured at a standard and at a two dimensional PGAA[3] setup (see also [can10]). As a reference we also conducted a chemical analysis with ICP-OES[4] and EDX measurements. The standard PGAA measurement found a Mo content of 7.65±0.24 wt% Mo in the U-8Mo target before sputtering, which is in fair agreement to the 9.2±0.6 wt% Mo determined by chemical analysis, and the 7.5±0.9 wt% Mo measured by EDX. All values indicate, that the value of 8.0 wt% Mo stated by the target manufacturer might be correct for the absolute Mo content of the target.

To identify inhomogeneities in the target bulk material or a shift in the overall composition, we conducted a two dimensional PGAA measurement of a target sample before and after sputter erosion [can11]. The two dimensional PGAA measurement provides here the advantage, that it allows mapping of the whole sample surface, like EDX analysis. In opposite to EDX analysis, that only measures several micrometers of the sample surface, PGAA gains information of the overall composition of the sample like a chemical analysis. Figure 4.2 shows the compositional variation for the non-used ('fresh') target sample, figure 4.3 for the used target sample.

It is clearly visible, that both samples show inhomogeneities in the spatial distribution of the constituent elements. The variation in the fresh sample amounts about 1.2 wt% and the inhomogeneous areas have a size in the range of 1 - 4 mm. They result apparently from the target production process. The measured composition of the U-8Mo with a Mo content of only 5.2 - 6.4 wt% also derivates significantly from the expected 8 wt% Mo. In the used sample we observed a variation of nearly 3 wt%. The Mo content varies between 4.1 - 7.0 wt%, and the size of the inhomogeneous areas is similar to the fresh sample. We have no evidence however, that the sputtering process might have caused the increase in

---

[3]Standard Prompt Gamma Activation Analysis (or: PGAA) uses a defined neutron beam to irradiate a sample of interest. The nuclear reactions induced by the neutrons in this sample produce a spectrum of prompt gamma rays, that can be measured. From the energies and intensities of this gamma spectrum it is possible to identify and quantify the elemental composition of the sample material. Two dimensional PGAA has the additional feature, that a sample can be scanned by the neutron beam, as it has a size of only 2 mm x 2 mm. This allows to create a 2D compositional map of the sample with the resolution given by the beam size.

[4]'Inductively coupled plasma optical emission spectrometry'

4 APPLICATION

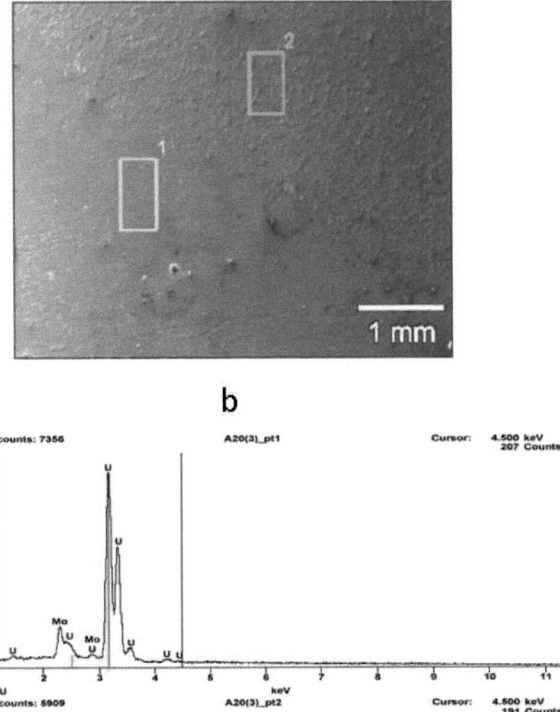

Figure 4.1: (a) SEM picture of the erosion zone on the used sputtering target. The labels show, where the composition has been measured by EDX. (b) EDX spectra recorded at the labeled areas.

U-Mo processing

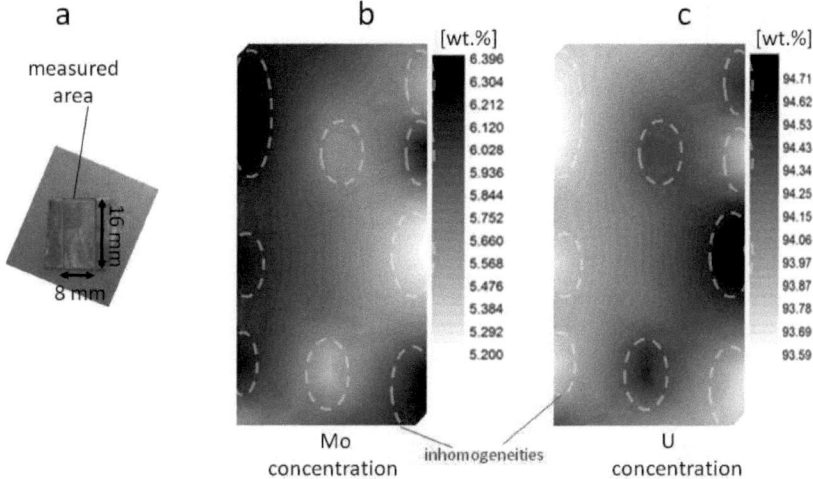

Figure 4.2: Material, that was cut out of a fresh U-8Mo sputtering target. A section of 16 mm x 8 mm has been mapped by 2D-PGAA. It is clearly visible, that the Mo is not distributed homogeneously inside the material. The inhomogeneity is more than 1 wt% and results from target manufacturing (figures from [can10]).

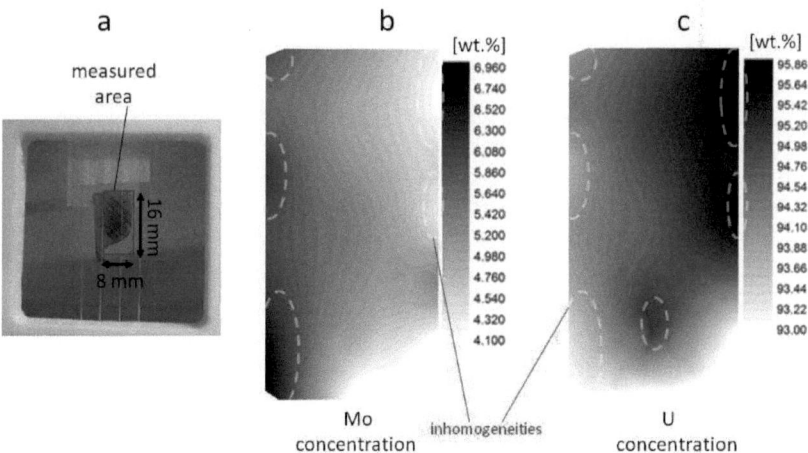

Figure 4.3: Material, that was cut out of a used U-8Mo sputtering target. A section of 16 mm x 8 mm has been mapped by 2D-PGAA. Again inhomogeneities are visible, this time however with nearly 3 wt% (figures from [can10]).

4 APPLICATION

the variation. We also do not have any evidence, that the total composition of the whole target is different from U-8Mo. Indeed we have clear evidence, that the Mo addition in the U-8Mo has generally a large spatial variation of at least 3 wt%, and that the local elemental composition can also significantly deviate from the expected composition.
Two dimensional PGAA also revealed the distribution of impurities in the U-8Mo. In the fresh target material we found impurities of Fe, Cu, Pb and V (see figure 4.4). The elements Fe, Pb and V seem to be arbitrarily distributed, while Cu can be found in the whole map. We thus suspect Cu to be a sample contamination, in particular because the fresh target samples had been cut by a wire erosion saw that used a brass wire. The observation of Hengstler [hen09] confirms this. In the used target material, only the impurity elements Fe and V could be identified. These elements are however not spread across the sample, but cover a large fraction of the whole map. Moreover, the distribution of Fe and V seems quite similar. Thus we suspect Fe and V to be contaminants in this sample. This would also be in accordance with the fact, that the used sample had been cut by a wire saw that used a diamond covered steel wire.

Film adhesion An aspect that has to be considered prior to U-Mo film deposition is, whether the deposited film should be free standing, that means it can easily be removed from the sputtering substrate, or if it should be tightly bond to the substrate after the process. In both cases the adhesion of the film to the substrate is of major importance, as in the first case it should be as low as possible and in the second case as high as possible.
As we wanted to have free standing U-8Mo deposits for our investigation, we decided to worsen the substrate adhesion. We thus deposited the U-8Mo on a Al substrate that had intentionally been covered with a thick oxide layer[5]. The bad adhesion between Al substrate and oxide film allowed us to remove the deposited U-8Mo films easily and by that to gain free standing films.

U-8Mo film deposition Figure 4.6 shows SEM pictures of the deposited U-Mo film. The surface is generally plain and homogeneous. At some points however some film defects are visible. From their irregular shape, the larger ones can be identified as particles, that dropped onto the film during sputter deposition (see figure 4.6a). In larger magnification it is also possible to identify organic contamination on the U-Mo film surface, that apparently had been applicated after the film was removed from the sputtering reactor.
Figure 4.7a shows a SEM picture of the smallest defects found on the films. The defect structures are quite symmetric and could result either from small dust

---
[5]This can easily be achieved if the Al is dipped into water.

U-Mo processing

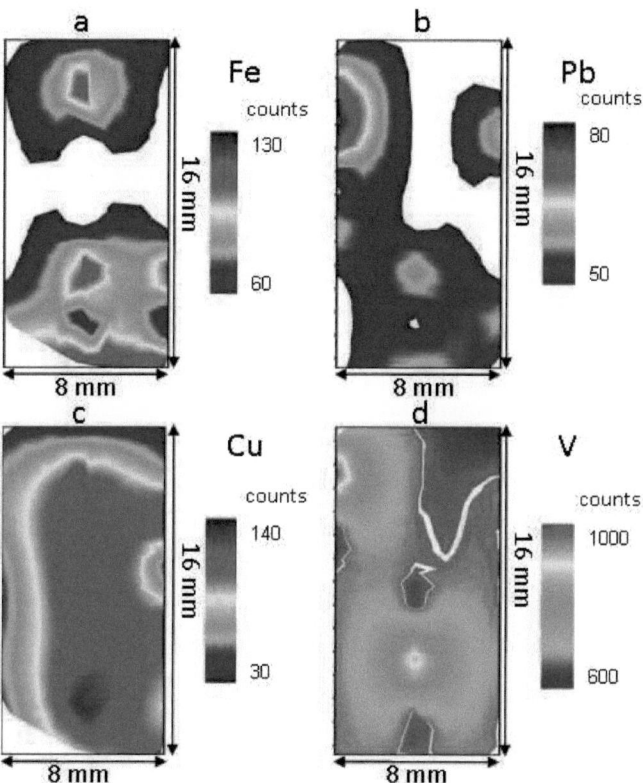

Figure 4.4: Impurities of Fe (a), Pb (b), Cu (c) and V (d), that had been found in the cross section of a U-8Mo target before sputtering (figures from [can10]). The shown concentration maps always refer to an area of measurement with the dimensions 16 mm x 8 mm. The concentrations are given in counts. While Fe, Pb and V seem to be spread inside the material, Cu is present in nearly the whole sample. This is easily explainable, as the samples for PGAA measurements were cut with a wire erosion saw that used a brass wire.

4 APPLICATION

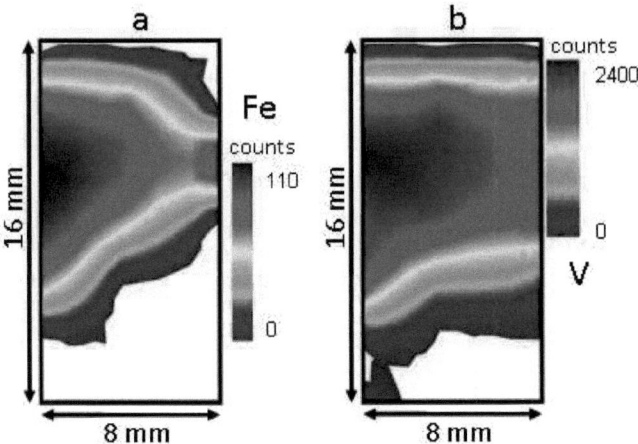

Figure 4.5: Impurities of Fe (a) and V (b), that had been found in the cross section of a U-8Mo target after sputtering (figures from [can10]). The shown concentration maps refer to an area of measurement with the dimensions 16 mm x 8 mm, concentrations are given in counts. The sample has been cut by a wire saw that used a diamond covered steel wire. We suppose measured Fe and V to be contaminations from this wire.

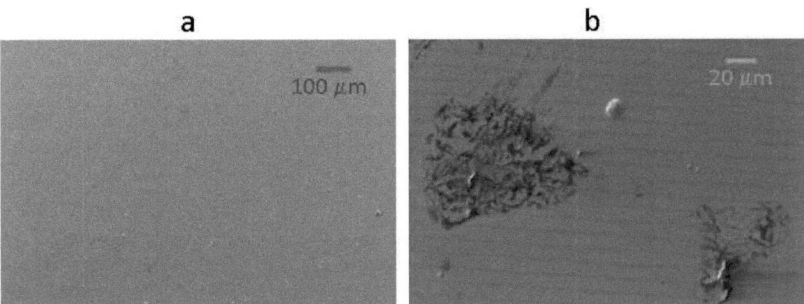

Figure 4.6: (a) SEM picture of a sputter deposited U-Mo film. Several minor defects are visible, that have a size of 10 μm or below. Most of them can be addressed to dust particles, that have fallen onto the film during growth. (b) Larger magnification of the SEM picture. The black structures were identified to be organic substances, that have contaminated the U-Mo film after removal from the sputtering reactor. The object in the upper center of the picture is also some dust particle.

U-Mo processing

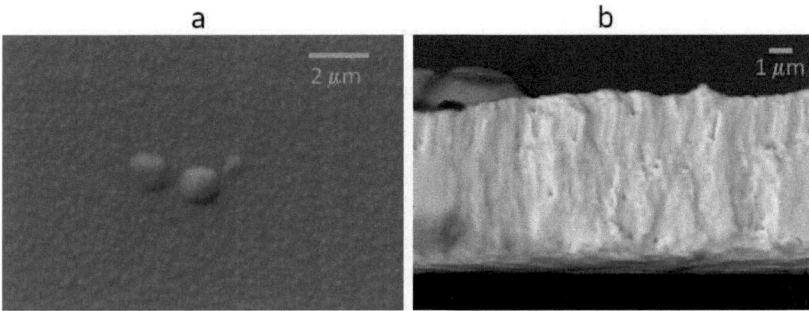

Figure 4.7: (a) SEM picture of the smallest visible defects found on the U-Mo film. These structures could either result from dust particles or from nodular defects. (b) SEM picture of a breaking edge of the U-Mo film. The densely packed fibrous grains, that are expected from the Thornton zone T, are clearly visible.

particles as well, but might also be nodular defects that evolved during film growth. These defects are not expected at a homologous temperature of 0.20 during the growth process, but could appear due to local surface roughness of the substrate. The structure of Thorntons zone T, that we expected for the U-8Mo deposit, can clearly be seen in figures 4.7a and b. Figure 4.7a gives a top view, figure 4.7b a side view of the densely packed fibrous grains the film is consisting of.

For comparison with the target material, we investigated the composition of the deposited U-Mo film as well. We used EDX for surface measurements, two dimensional PGAA for compositional mapping as well as XRD to determine the crystal structure of the deposit. The EDX measurement gave us a Mo content of 7.52 ± 0.08 wt% at the film surface. This is identical to the composition of the U-8Mo target within the accuracy of measurement.

We wanted to use two dimensional PGAA to get a composition map of the U-Mo film. A complete deposit of 100 mm x 100 mm size was however too large for the PGAA facility, hence we decided to use only one piece of the film sized 50 mm x 50 mm. Two PGAA line scans were conducted in the middle of the film, one in horizontal, one in vertical direction (see figure 4.8). The area of measurement was each time 50 mm x 2 mm. The measurement gave the unexpected result, that for both horizontal and vertical scan, the Mo concentration seems to show a minimum in the middle of the scan region and a maximum at the edges. For the vertical scan, the Mo concentration was determined to 8.2 - 10.1 wt%, for the horizontal scan, we measured 7.4 - 9.8 wt%. If the position of the measured piece of film during deposition is regarded, it seems possible, that this might

4 APPLICATION

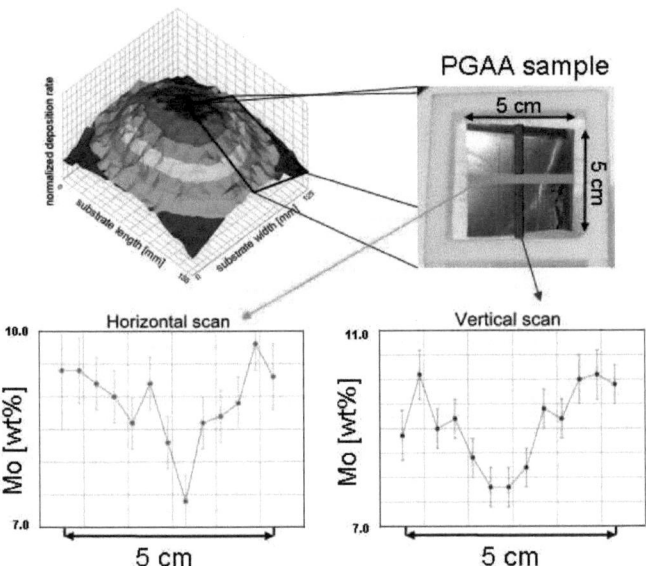

Figure 4.8: A 100 mm x 100 mm Zr substrate was coated with 50 μm of U-8Mo and cut into four pieces of 50 mm x 50 mm size. One of these pieces was measured by two dimensional PGAA in horizontal and in vertical direction. It turned out, that the concentration of Mo reaches a minimum in the middle for both scan directions. This indicates a ring shaped area of increased U concentration, that possibly has been caused by different ejection characteristics of U and Mo in connection with the ring geometry of the plasma.

indicate a difference in the ejection distribution of U and Mo during sputtering. As the plasma configuration used for sputtering shows a ring shape, it can be assumed, that different ejection characteristics of U and Mo would also result in ring shaped areas of different U and Mo composition on the whole deposited film. A line scan conducted on a quarter piece of such film would measure a quarter of these ring structrues, and should thus be symmetric in horizontal and vertical direction. The similarity in the Mo concentration distribution recorded for the horizontal and vertical scan direction suggests such ring structure, or, more precisely, a ring area of significantly increased U content.

To investigate the phase of the U-Mo film, we conducted several XRD measurements [hen08]. We found, that the deposited film has a bcc crystal structure (space group Im-3m), which could be identified as the γ-phase of U-Mo that we expected to gain at room temperature (see also appendix A3). The Rietveld analysis also confirmed a strong preferred orientation in the [110] direction, which of course results from the Thornton zone T fibrous grain structure in the film.

4.1.2 Surface cleaning

Sputter erosion can be used to remove any type and thickness of surface pollution from U-Mo. We investigated the surface cleaning effect at the example of oxidized monolithic U-10Mo foils. Their surface was analyzed before and after sputter erosion.

U-Mo surface pollution For U-Mo the surface cleaning is more difficult than for most other materials, as U has a very high affinity to oxygen and thus a high rate of oxidation under air. Moreover U-Mo does not form a protective oxide barrier like many other metals, which means, that the thickness of the oxide layer increases continuously over time. Figure 4.9a shows the SEM picture of a U-10Mo substrate before cleaning. The substrate is covered with an oxide layer of about 10 μm thickness.

The sputter erosion cleaning of an oxide layer like this would take several ten hours in our sputtering reactors[6]. In practice, it is therefore much faster and more convenient to perform a rough cleaning step with chemical or mechanical aids prior to the actual sputter erosion cleaning. In this combination the sputter erosion acts as a final in-depth cleaning. By mechanical grinding or chemical baths, it is possible to remove more than 99% of the oxide layer in several minutes (see figure 4.9b). The thin oxide film remaining on the surface has a thickness in the nanometer scale, but is still perturbing for bonding purposes and a possible source of bonding problems. Moreover chemical baths can leave residues on the fuel foil surface, that are generally undesired as they impose the danger of local corrosion to fuel and cladding. Films in the nanometer scale consisting of oxides and chemical residues can however be removed quickly and efficiently from the surface by sputter erosion.

U-10Mo surface cleaning We prepared three monolithic U-10Mo foil samples, that all had a surface that was completely oxidized (see figure 4.9a). Each foil had to be roughly cleaned by grinding (see figure 4.9b) before we started with

---

[6]Actually, the sputter erosion process in our sputtering reactors is much slower than the sputter deposition process, as the sputter erosion uses the substrate as sputtering cathode and the target as sputtering anode. The sample table respectively the carrier electrode are however not equipped with a magnetic assembly, and there is no magnetical electron confinement to support the glow discharge plasma. Thus the substrate erosion process can only be operated in a normal glow discharge mode. This operational mode has the disadvantage, that it has much lower erosion rates than the magnetically enhanced mode, but also the advantage, that the erosion will be homogeneous on the whole substrate surface.

4 APPLICATION

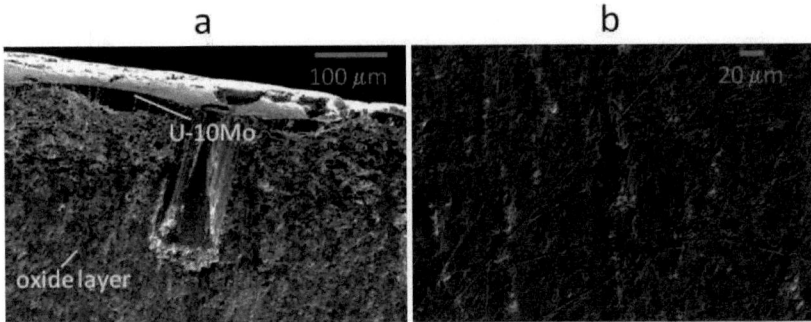

Figure 4.9: (a) SEM picture of surface and cutting edge of a U-10Mo foil sample before cleaning. The substrate is covered with an oxide layer of about 10 μm thickness. (b) SEM picture of surface after mechanical cleaning (grinding and polishing). The scratches of the cleaning process are clearly visible, the oxide layer has nearly completely been removed.

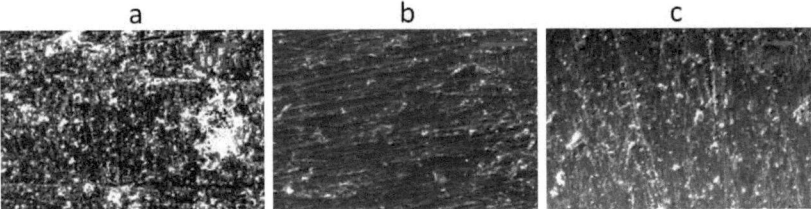

Figure 4.10: (a) SEM picture of the surface of a U-10Mo foil sample that was mechanically grinded. The surface is covered with scratches and remaining oxide clusters (white). (b) SEM picture of the surface of a U-10Mo foil sample after grinding and 10 minutes of sputter erosion at 100 W. The scratches are still visible, but the oxide clusters have mainly disappeared. (c) SEM picture of the surface of a U-10Mo foil sample after grinding and one hour of sputter erosion at 100 W. There is hardly any visible difference from the situation in picture (b).

our cleaning experiments. The grinding was done in air and without chemical aids. After grinding, the foils were inserted and stored in Ar atmosphere to avoid further oxidation before the sputter erosion cleaning.

The cleaning process was conducted in the full size sputtering reactor. The polarity of the sputtering diode was switched before, so that the carrier electrode was the cathode. In this configuration, the plasma in the reactor is slightly unstable and shows the tendency of arcing. It could be ignited at a working gas pressure of $6 \cdot 10^{-3}$ mbar and was operated at 100 - 300 W (see figure 4.16b). We however avoided going to much higher plasma powers, as this would both increase the tendency of arcing and also make the appearing arcs more powerful. We operated the reversed sputtering process for time scales from several minutes up to more than one hour.

Figure 4.10a shows the resulting surface after mechanical grinding. The surface is dominated by scratches from the grinding and oxide clusters that have not been removed. Figures 4.10b and c show pieces of the same foil that were sputter erosion cleaned after the grinding. Even 10 minutes of cleaning at only 100 W significantly reduces the amount of oxides remaining on the surface. One hour of sputter erosion has the same effect on surface cleanliness, but apparently no additional gain.

4.1.3 Coating

The application of coatings by sputter deposition is the most common use of sputtering in industry. There are many different kinds of coatings for various purposes presented in literature. We investigated the surface coating of monolithic U-10Mo foils as well as the most crucial property of the coating, the substrate adhesion.

U-10Mo surface coating In principle the coating of a U-Mo surface is only a film formation process as it was described in 4.1.1, with the only difference that U-Mo is now the substrate and other materials are used as target. We deposited coatings of several different materials onto monolithic U-10Mo foil samples using the tabletop as well as the full size reactor. In both setups we conducted the sputter deposition process at pressures in the range of $3 \cdot 10^{-3}$ mbar at plasma powers of 240 W respectively 2 kW. Each foil sample had been roughly cleaned by grinding and afterwards cleaned by sputter erosion. The sputtering process was operated over time scales of 1 - 10 h, and coatings in a thickness range of 2 - 25 μm were realized. Figure 4.11 shows two examples of coated U-Mo with different coating materials in different thicknesses.

In contrary to the previous case, where free standing films were of interest and the substrate to film adhesion was intentionally worsened, it is demanded for a coating to show a good substrate to film adhesion. We investigated the quality of adhesion by tensile tests.

Coating adhesion We conducted tensile tests to investigate the adhesion of different coatings on U-Mo substrates respectively of U-Mo films on different substrates. Details on the basics of tensile tests are described in appendix A4.
We tried to eliminate all disturbing factors in the tensile tests by choosing our sample geometry to feature a planar interface between the different materials. Moreover we wanted to test the adhesion of only two materials to each other

4   APPLICATION

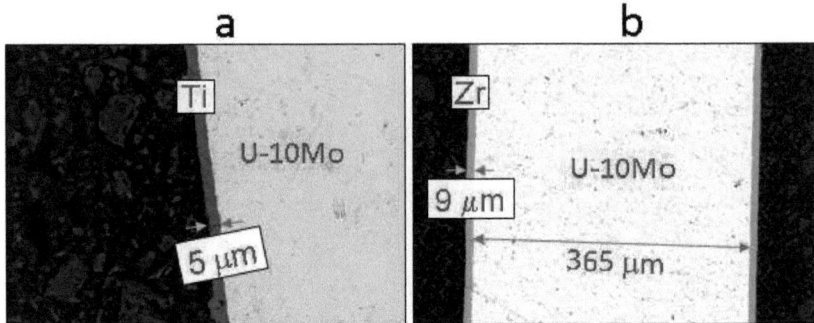

Figure 4.11: (a) Cross section of a U-10Mo foil sample coated with 5 μm of Ti. (b) Cross section of a U-10Mo foil sample coated with 9 μm of Zr on both sides.

substrate	U-8Mo	Ti	Zr	Zry-4	Nb
U-10Mo		15	15	15	15
Ti	15				
Zr	15				
Zry-4	15				
Nb	15				
Ta	15				
Al	15-20	15	15	15	15

Table 4.1: Material combinations of coating and film adhesion tests. The left column lists the different substrates that have been used, the first line lists the deposited material, the number denotes the thickness of the sputtered deposit in μm.

at a time[7]. Figure 4.12a shows the principle setup of our tensile test samples. Each sample consists of two layers, the substrate and the deposited film, that are bonded to each other only by the film growth process. The substrates with a thickness of 100 - 400 μm are polished respectively grinded foil samples, according to the surface quality that is supposed to be investigated. The films to be tested are sputter deposited on these blank sheet pieces in a thickness of 15 - 20 μm. The two layer sample created in this way is glued between two sample holders with a special adhesive. After hardening of the adhesive, the sample is ready for a tensile test (see figure 4.12b).

We produced a series of different tensile test samples in the described method [dir10] [jur11]. Table 4.1 shows the material combinations that have been realized.

---

[7]We did not consider three layer systems, as these systems present two interfaces and therefore two different adhesion values.

U-Mo processing

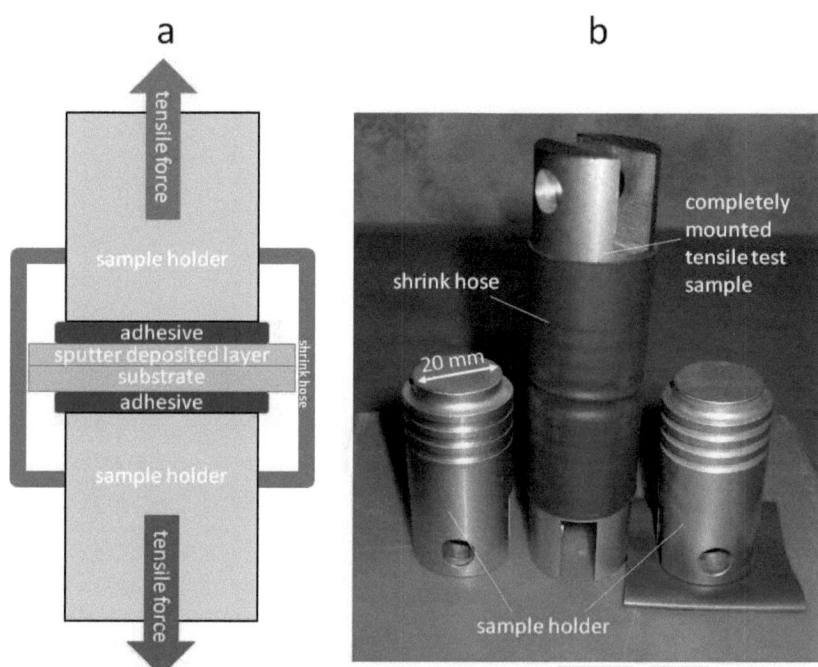

Figure 4.12: (a) Basic setup of the used tensile test samples. A substrate that was sputter coated with another material is glued between two sample holders with a special adhesive. After that the tensile test sample is enclosed into a shrinking hose, that retains radioactive particles that may eventually appear during the sample fracture. (b) Picture of an actual tensile test sample, that was completely mounted and is ready for testing (middle), as well as two of the sample holders used (left and right) (from [dir10]).

105

4 APPLICATION

combination	adhesion [MPa]
U-Mo / Nb	> 70
U-Mo / Ta	> 62
U-Mo / Ti	39 - 70
U-Mo / Al (polished)	32 - 53
U-Mo / Al	18 - 51
U-Mo / Zr	67 - 69
U-Mo / Zry-4 (polished)	22 - 39
Al / Nb	55 - 56
Al / Ti	40 - 68
Al / Zry-4	> 24

Table 4.2: Adhesion values of the different combinations measured by Dirndorfer [dir10].

Each one was mounted into a tensile test machine and an increasing tensile force normal to the interface of the two sample materials was applied until a fracture occurred. It turned out, that the breaking of the samples did not always occur due to a layer separation of the tested materials. In most cases the used adhesive lost contact and separated either from the sample or from the sample holder. Because of that, all tensile test samples were investigated after the tensile test and classified into different breaking classes. Dirndorfer [dir10] distinguished between three classes: the separation of the sputter deposited layers, the separation of the adhesive from one of the contacted surfaces and the breaking of the adhesive in itself. From the first one of these breaking classes an actual breaking strength value of the deposited samples could be deducted, the second and third class only gives a lower limit of the breaking strength. Table 4.2 lists the results he gained for the breaking strength of our sputter deposited samples.

The adhesion of sputter deposited coatings is in all cases better than 18 MPa, usually even above 30 MPa. This is an excellent value compared to other bonding methods [cla06]: the maximum adhesion of material bonded by friction stir welding (or: 'FSW') process is given as only 6.42 MPa, the adhesion of transient liquid phase bonded (or: 'TLPB') material given as 15.4 MPa. Only the hot isostatic pressing process provides an adhesion of up to 60.3 MPa, which is comparable to the better fraction of adhesion values measured for our coatings. It should however be noted, that these values are only the values we reached with our current sample preparation method, not the maximum possible adhesion values.

Application I: Fuel fabrication

Figure 4.13: (a) Hot rolling assembly for monolithic U-Mo foil production (from [moo08]). Cast U-Mo coupons are laminated into a carbon steel picture-frame assembly. The assembly is 'canned', by welding of bottom and cover plates, and hot rolled after that. (b) After hot rolling the assembly is decanned, and the rolled U-Mo foil sheared to the desired size (from [moo08]).

4.2 Application I: Fuel fabrication

Three uses of U-Mo processing in the field of fuel fabrication seem reasonable: the formation of a massive U-Mo fuel foil, the application of diffusion preventive coatings onto U-Mo fuel and the application of cladding. We investigated all of them.

4.2.1 Fuel foil

Techniques to fabricate monolithic U-Mo fuel foils have been developed in the past decade with great effort and are available since some years [moo10]. Currently these techniques are operative on laboratory scale only, and several years of development will still be necessary until an industrial scale production seems actually feasible.

State of the art The only working U-Mo foil production process up to now is described in [moo10]. Cast U-Mo coupons are laminated into a carbon steel picture-frame assembly (see figure 4.13a).

Bottom and cover plates from carbon steel are applied and welded onto the assembly. The whole process is conducted in Ar atmosphere. The resulting 'canned' structure is heated and rolled several times, which can be performed in air as the insulation canning protects the U-Mo from oxidation. The sequence of heating and rolling is repeated until the desired assembly thickness is reached. After rolling, the assemblies are annealed, mechanically decanned and sheared to the desired size. The result is a U-Mo foil as shown in figure 4.13b.

4 APPLICATION

The described procedure is mainly processed manually, and up to now only suited for the production of smaller numbers of fuel foils [tec08]. Without a higher degree of automation, an industrial production seems hardly feasible.

Sputter deposition Böni and Wieschalla suggested the fabrication of monolithic U-Mo fuel foils by sputter deposition as an alternative fabrication technique. According to their patent from 2006 [pat06], this method could be superior to the currently existing production process. The suggested method was to use sputter deposition to grow a massive U-Mo film, that is in material, microstructure and dimension identical to a conventionally produced monolithic U-Mo fuel foil. The idea seems promising, and basically it should be simple to realize in our sputtering reactors.

Material and microstructure The exact composition of the U-Mo alloy used for a monolithic fuel foil, that means the exact fraction of Mo in the U, is determined by metallurgical considerations and by the particular parameters of foil fabrication and processing. A necessary lower limit of the Mo fraction can be set at about 4.5 wt%, as this value is needed to stabilize the U γ-phase (see appendix A3). A necessary upper limit for the Mo content is 15.5 wt%. We chose the alloy U-8Mo for our sputter deposition experiments, as it seemed us to be a fair compromise between phase stability and U density. Several U-8Mo sputtering targets as shown in figure 3.21 were thus purchased from AREVA-CERCA.

The desired crystallographic structure in a monolithic U-Mo fuel foil is the body-centered cubic γ-phase, that shows an isotropic thermal expansion behavior (see also appendix A3). When a fuel foil is produced by sputter deposition, there are two ways to obtain the γ-phase (see figure 3.26): a deposition at low homologous temperature to immediately receive a 'frozen' γ-phase or a deposition at a high homologous temperature and a quick cooling of the resulting film[8]. We decided to use the first option, as a fast film cooling in the full size reactor seemed us to be much more difficult and risky.

Foil dimensions The required dimensions of a monolithic U-Mo fuel foil is dependent on the reactor it will be used in. The 'Advanced Test Reactor' (or: ATR) at the Idaho National Laboratory would for example require monolithic U-Mo foils in the size 600 mm x 60 mm x 360 μm, the FRM II would require foil dimensions of 700 mm x 62.4 mm x 425 μm[9].

---

[8] According to Tangri [tan61], a cooling rate of 10 °C per second should be able to conserve the γ-phase in U-8Mo.

[9] Breitkreutz [bre11] determined a necessary thickness of approx. 425 μm, if the fuel alloy U-8Mo would be used.

Application I: Fuel fabrication

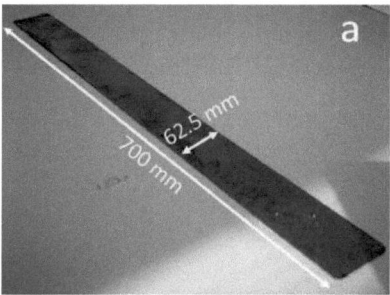
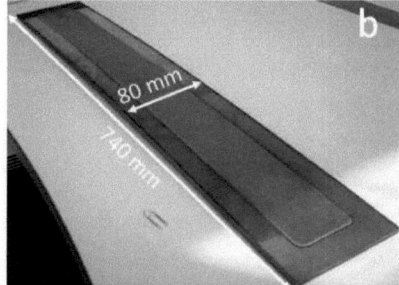

Figure 4.14: (a) Self sustaining Cu film with dimensions 700 mm x 62.4 mm x 1200 μm. The structure was deposited onto a strongly oxidized Al surface and could therefore be removed from it. (b) Cu deposit with dimensions 700 mm x 62.4 mm x 425 μm deposited into an Al plate with a machined pocket. The adhesion between Cu and Al is excellent, and causes a slight curvature of the plate due to the bimetal effect (see also chapter 2.1.6). During deposition the Al plate was covered by a pattern that shielded the machined pocket. Thus there was only material deposition into the pocket.

In a test with Cu as surrogate material, we investigated how much effort would be necessary to deposit a foil of such dimensions. In about 42 continuous hours of sputtering at a power of 4 kW we were able to produce a free standing Cu film with dimensions 700 mm x 62.4 mm x 1200 μm in the full size sputtering reactor, which easily meets the dimensional requirements mentioned before (see figure 4.14a). We were also able to deposit a Cu film with dimensions 700 mm x 62.4 mm x 425 μm into an Al plate with a machined pocket (figure 4.14b). The pocket design was considered an alternative to a free standing deposit.

Our surrogate tests thus showed, that also the sputter deposition of a U-Mo film in the dimensions required for a fuel foil should be feasible in the full size reactor. Subsequently we were supposed to continue our experiments with the deposition of a U-8Mo fuel foil in full size. Considerations about the principal material efficiency of the sputter deposition process and the scientific insight, that could be gained by the deposition of a full size fuel foil, however changed our further approach.

Efficiency   During our experiments it soon became clear, that the sputter deposition process as we use it in our reactors has only a low degree of material utilization. This results both from our plasma geometry as well as from the reactor geometry.

The ring shaped plasma geometry, that we use in our sputtering reactors, has the advantage to increase deposition speed significantly and to allow the operation of a glow discharge plasma at much lower pressures as it would normally be possible. However it has also the disadvantage, that deposition in the reactor is not

4 APPLICATION

homogeneously any more, and that sputter erosion takes place only at defined erosion zones (see figure 3.23). One effect of our plasma geometry is hence, that the used sputtering targets are not totally consumed during the erosion process, but only to a certain fraction. For the brick-shaped target elements that we use in both our sputtering reactors, we determined a target utilization of only 20 - 25 % for the tabletop reactor and 30 - 35 % for the full size reactor, which means that only this weight fraction of the target material is sputtered and ejected from the target, while the remaining target material mass stays unused[10]. Moreover will most of the material, that is sputtered and ejected from the target, be deposited to the walls of the reactor and not onto the substrate. This effect results from the reactor geometry, i.e. more precisely from the shape of sputtering target and substrate and their assembly, and further lowers the efficiency of the process. In our reactors, we determined that only about 15 - 25 % of the ejected material is deposited onto the substrates, depending on the size of the substrate.

Both factors, the target utilization and the fraction of correctly deposited material, set the degree of material utilization in our sputtering reactors to 3 - 6 % for the tabletop reactor and to 4 - 9 % for the full size reactor. In case of a cheap target material like Al, the material utilization factor can be ignored, and only the effort in time and energy of the sputter deposition process has to be considered. In case of an expensive target material like U-Mo, a low utilization factor is however a major disadvantage and fatal for efficiency and competitiveness of the process. In comparison with the currently used process of U-Mo foil fabrication, that has a degree of material utilization of around 76 % (see section 4.2.4), the processing of U-Mo by sputter deposition is thus inefficient and not competitive.

In the particular case of depositing a monolithic U-8Mo foil in full size we estimated the effort as follows: an U-8Mo target for the full size reactor consists of 13 single U-8Mo target bricks, that have a total mass of 14.8 kg. Assuming the mentioned material utilization of 4 - 9 % for the full size reactor, about 3 - 7 full sized monolithic U-8Mo foils could thus be produced from these 13 U-Mo target bricks, and at least about 13.5 kg of U-8Mo scrap would remain. In the view of these numbers, and considering the expensive U-8Mo target material[11], we regarded the gain of scientific information by a foil production like this as limited and dispensable. Nevertheless we conducted a series of experiments to investigate the deposition behavior of U-8Mo in our tabletop reactor.

Deposition of monolithic U-8Mo We deposited U-8Mo films of various dimensions and thicknesses on various substrates (see figure 4.15). We reached thicknesses of up to 120 μm, but usually stayed below 50 μm to save material. The

---

[10] Values like that are common for magnetron sputtering reactors [fre87].

[11] One single U-8Mo target brick from DU with a mass of approx. 1.1 kg costs approx. 5000 euro. A comparable U-Mo target from LEU or HEU is assumed to cost significantly more.

substrate size was limited to 100 mm x 100 mm in the maximum, as larger substrates could not be placed inside the tabletop reactor. The deposition was usually conducted at a plasma power of 240 W, a working gas pressure of $3 \cdot 10^{-3}$ mbar and a substrate temperature of 15°C. As expected, the material was always found to be in the γ-phase.

Conclusion We could demonstrate, that the idea of sputter depositing a complete U-Mo fuel foil is in principally feasible. The process is however time- and energy-intensive, as several ten or even hundred hours of sputtering at plasma powers in the kW range would be necessary to reach the necessary thicknesses. The material utilization in our full size sputtering reactor was moreover only in the range of 4 - 9 %, which is fatal, when an expensive material like U-Mo has to be processed. Even if this utilization rate could be increased it is questionable, whether the sputter deposition of U-Mo fuel foils would in general be competitive to the current foil fabrication method.

4.2.2 Barrier coating

The application of functional coatings for nuclear fuels is investigated since decades, mostly to apply neutron poisons to the fuel itself [pat86][pat91] or to increase the wear resistance of the cladding [pat93]. In the case of U-Mo fuels, the application of functional coatings is however still in its early stage. Currently, the only purpose in discussion why functional coatings should be applied to U-Mo fuels, is as a barrier to prevent an IDL formation between fuel and cladding. We therefore concentrated on these diffusion preventive coatings. Other possible uses seem evident[12], but have not been studied up to now.

Diffusion prevention IDL formation is caused by a radiation enhanced thermal diffusion process (see next section). Two approaches to prevent or slow down IDL formation involving fuel coatings are possible. The first approach tries to immobilize the reacting atoms. The fuel coating in this case is very thin (submicrometer), and has the purpose to localize appropriate reactive materials at the interface between U-Mo and Al. During reactor operation, this kind of coating is supposed to be consumed by interaction with U-Mo or Al. The immobile compound that forms on the U-Mo/Al interface in this interaction then blocks further diffusion and prevents IDL formation. The second approach tries to stop the fission products before they can actually reach the Al. To guarantee that, the applied

---

[12]Coatings of appropriate materials could avoid oxidation of the fuel surface during fabrication, could simplify the bonding process of fuel and cladding or provide wear resistance during fuel handling. Moreover could neutron poisons be placed close to the fuel to retain surplus reactivity.

4 APPLICATION

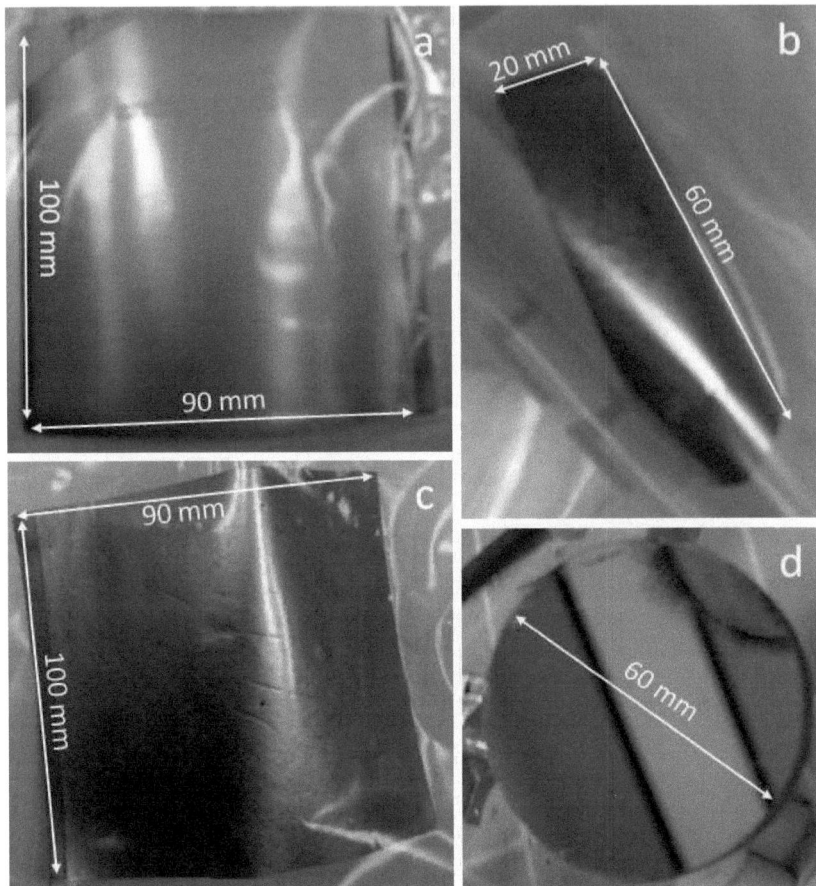

Figure 4.15: U-8Mo deposits produced in the tabletop sputtering reactor. (a) Deposit of dimension 100 mm x 90 mm x 50 μm on a 50 μm tantalum foil. The substrate adhesion was excellent and the film showed no sign of delamination. (b) Free standing deposit with a dimension of 60 mm x 20 mm x 10 μm. (c) Deposit of dimension 100 mm x 90 mm x 50 μm on a 50 μm graphite foil. The substrate had too less mechanical stability to withstand the stresses imposed by the U-8Mo film and was thus deformed already during deposition. The substrate adhesion was good, but by mechanical force it was possible to remove the film from the surface as the substrate could not resist. (d) Deposit of 60 mm diameter and 25 μm thickness on glass. The substrate adhesion was perfect if the surface was prepared properly.

Application I: Fuel fabrication

coating has to be thicker than the maximum range of the fission products emitted from the U-Mo, that means typically about 10 - 20 μm. The coating thus avoids, that the area that would allow radiation enhanced thermal diffusion reaches the Al, and by that prevents an IDL formation between U-Mo and Al. Instead, the coating will be in the area of radiation enhanced thermal diffusion, and the coating material has of course to be chosen appropriately.
We performed a material pre-selection, that identified UO_2, Zr, Ti, Nb and Ta as suitable materials for a diffusion preventive coating [jar09]. Coatings and additions of Si [per09] and ZrN [bir06] [bir09] have also proven to reduce IDL formation and are therefore of relevance to us.

State of the art The currently used technique for the coating of monolithic U-Mo fuel is colamination [moo08] (see also chapter 1). Colamination means, that the U-Mo coupons used in the foil fabrication procedure (see section 4.2.1) are sandwiched with Zr blank sheets prior to the lamination into the carbon steel picture-frame assembly. The following hot rolling sequence will thus not only reduce the thickness of the U-Mo coupon, but also bond it to the Zr. The result is a U-Mo foil coated with Zr.
As the foil production and the coating are done in one single process step, the colamination technique suffers from the same weaknesses as the foil production. Again sputter deposition seems to be a promising alternative, as it is widely used to apply coatings in different fields.

U-10Mo fuel coating Full size U-10Mo foils were provided to us by courtesy of Y-12 National Security Complex. Each foil was approximately 600 mm x 60 mm in size and 360 μm in thickness (see figure 4.16a). One of the foils was cut into smaller pieces with a size of approximately 60 mm x 30 mm each. These pieces will be denoted as minifoils.

The surface of each foil respectively minifoil had to be roughly cleaned by grinding before we started with our experiments. After grinding, the foils were inserted into the full size reactor. As next processing step, we used sputter erosion to achieve an in-depth cleaning of the surface of the U-Mo foils (see figure 4.16b). This cleaning process was maintained for approximately 10 - 15 minutes at a sputtering power of 100 W. After this cleaning step, the coating of the foils started (see figure 4.16c). The coating process was maintained for several hours at a plasma power of 1 kW until the desired layer thickness had been reached.
By the described process, the configurations listed in table 4.3 have been realized.
Minifoil coatings with Zr and Zry-4 were performed prior to the full size coatings, to clarify whether a stable coating upon a grinded surface can successfully be

4 APPLICATION

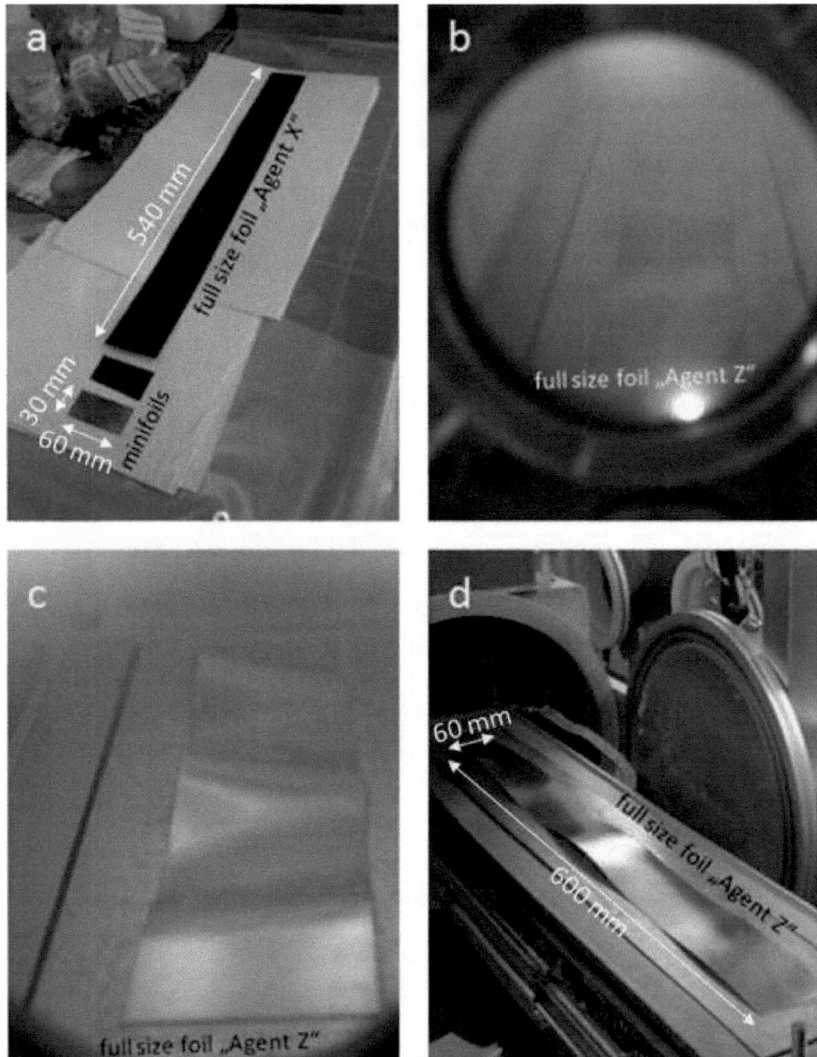

Figure 4.16: (a) The full size foil 'Agent X' was cut into several minifoils, which were used for the first coating tests. (b) The full size foil 'Agent Z' during the sputter cleaning and (c) during the sputter coating process. The ripples in the foil do not result from cleaning or sputtering, but were already present in the foil when delivered. (d) 'Agent Z' after the coating with Zry-4 and AlFeNi.

Application I: Fuel fabrication

number and description	size [mm x mm]	coated with
2 minifoils (from Agent X)	28 x 30	Zr, Zry-4
6 minifoils (from Agent X)	60 x 30	Zry-4
1 full size foil (Agent Z)	600 x 60	Zry-4
1 full size foil (INL 4 - 2A)	600 x 60	Ti

Table 4.3: U-10Mo minifoils and full size foils that have been coated by sputter deposition.

achieved. This was not self-evident, as all coating tests before had used polished substrates, but the resulting coatings turned out to be stable indeed. After that we coated one full size foil with Zry-4 and another one with Ti. All sputter coated foils and minifoils listed in table 4.3 optically appeared excellent and don't show any signs of delaminating (see figure 4.11). The tensile tests described in the previous section confirmed the excellent coating adhesion.

Conclusion We demonstrated, that small and full size U-Mo fuel foils can easily be coated with the diffusion preventive materials Zr respectively Zry-4 and Ti in a thickness of 1 - 25 μm within several hours of sputter deposition at plasma powers in the kW range. The adhesion of these coatings is according to tensile test experiments comparable to coatings produced by conventional techniques and usually above 30 MPa [dir10]. We expect similar results for other diffusion preventive materials like Nb and Si, as the deposition of these materials has already been successfully demonstrated at small samples for irradiation tests (see section 4.3). The sputter coating technique thus seems to be ideally suited to apply diffusion preventive materials onto monolithic U-Mo fuel. The combination of the sputter erosion cleaning of U-Mo foil surfaces with the sputter coating of these foils provides a quality of surface preparation that cannot be reached by the techniques currently used in U-Mo fuel fabrication. For this type of application the sputtering technique can thus be considered as clearly superior. An integration into the currently fuel fabrication process seems to be advisable and feasible.

4.2.3 Cladding

The last step in monolithic fuel plate assembly is the application of cladding to the fuel. The bonding of fuel to cladding is a crucial point during plate assembly as a fuel plate can only be used, if an appropriate bonding is assured.

State of the art To assemble a monolithic fuel plate a U-Mo fuel foil and two Al cladding plates are needed. A 'sandwich' of Al bottom cladding plate, fuel foil surrounded by Al frame and Al cover cladding plate is assembled by the

4   APPLICATION

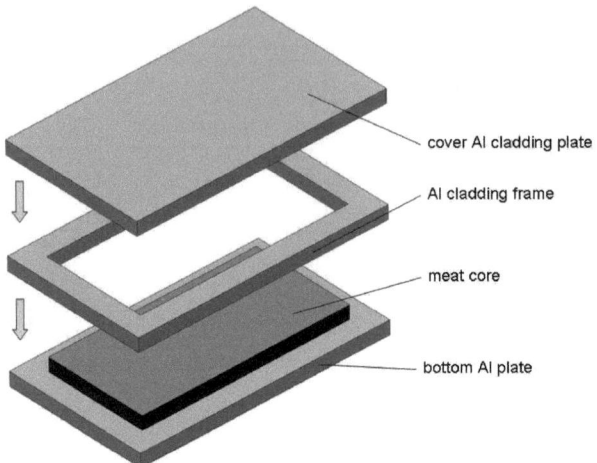

Figure 4.17: Sandwich assembly of a fuel plate. The meat core is framed by a border of Al cladding. A top and bottom Al cladding plate are applied, then the cladding has to be bonded.

well-known picture-frame technique (see figure 4.17). The sandwich is bonded in a so-called 'friction bonding' (or: FB) process using a rotating pin tool that is pressed onto the sandwich with force [moo08]. As alternative techniques to FB the so-called 'hot isostatic pressing' (or: HIP) technique and the hot rolling technique are also investigated, where the sandwich is pressed for several hours at constant pressure and at elevated temperatures respectively rolled at elevated temperatures. The result in all three cases is a complete fuel plate.

Böni and Wieschalla also proposed sputter deposition as an alternative technique for this application [pat06]. The idea is promising, and was thus investigated by us.

Cladding   The application of cladding to a monolithic U-Mo fuel foil is quite similar to the application of a functional coating discussed in the previous subsection. The relevant cladding materials are Al and Al alloys, usually AlFeNi and Al-6061. Only the thickness of the cladding layer is in the range of several hundred micrometers and therefore much larger than the thickness of a functional coating.

We conducted several sputter deposition tests to clad full size Fe and Cu surrogate foils with the mentioned materials. Chemical cleaning and sputter erosion cleaning were used before the deposition process to ensure the necessary surface cleanliness of the foils. After that, the foils were cladded with different layer

Application I: Fuel fabrication

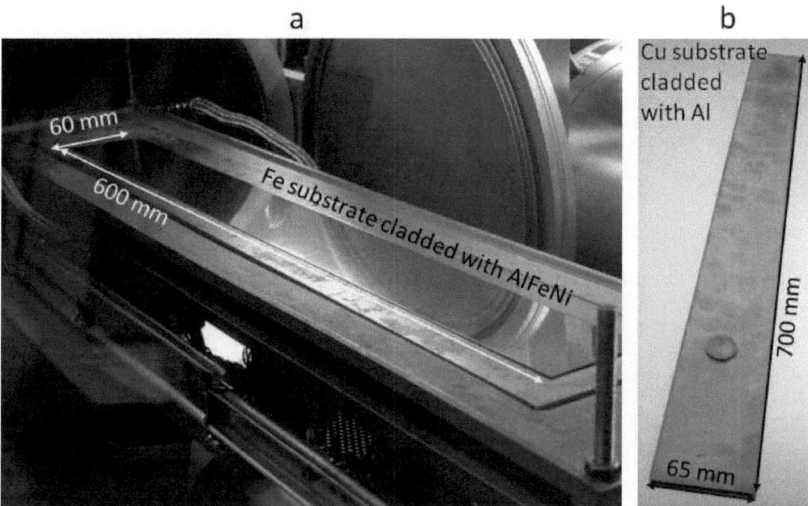

Figure 4.18: (a) Fe surrogate foil cladded by 100 μm of AlFeNi. (b) Cu surrogate foil cladded by 50 μm of Al.

thicknesses from 1 - 100 μm. Figure 4.18 shows a Fe and a Cu surrogate foil, that have been cladded with AlFeNi respectively Al.

Again, the cladding procedure was simple and the resulting cladding showed no signs of delamination. Several ten hours of sputter coating were however necessary for each foil to reach the mentioned thicknesses. We did not conduct any experiment to go beyond 100 μm cladding thickness, as this value did not seem reasonable to us (see following section).

As already mentioned, the fuel foil in a monolithic fuel plate for FRM II is supposed to have dimensions of 700 mm x 62.4 mm x 425 μm [bre11]. The complete fuel plate is fixed to the outer dimensions of 720 mm x 76 mm and a thickness of 1360 μm. The AlFeNi cladding thus has to have a thickness of approx. 470 μm on each side of the fuel foil. Moreover is has to be in total 20 mm longer and 13.6 mm wider than the fuel foil itself.

A cladding thickness value of approx. 470 μm can be reached by several ten hours of sputter deposition, which makes this technique comparatively slow compared to the picture-frame technique. A huge advantage of a sputter deposited cladding is however the excellent adhesion between fuel substrate and cladding film. What cannot be satisfyingly reached by sputter deposition is the length and width of the cladding, as it is not possible to deposit a cladding layer that is in width and length larger as the underlying fuel foil without creating an undesired step pro-

4 APPLICATION

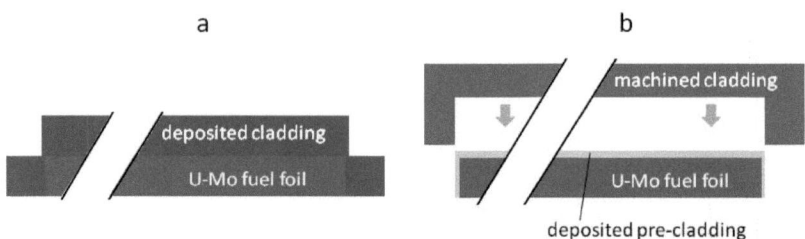

Figure 4.19: (a) Cladding application by sputter deposition. A problem occurs at the foil edges: it is not possible to deposit a cladding layer that is in width and length larger as the fuel foil without creating an undesired step profile in the total thickness of the resulting fuel plate. (b) Pre-cladding concept: a thin layer of cladding material with several micrometers thickness is applied to the fuel foil by sputter deposition. This 'pre-cladding' simplifies bonding to the actual cladding. The pre-cladded fuel foil is laminated into a massive cladding plate with machined pocket, and can easily be roll bonded afterward.

file in the total thickness of the resulting fuel plate. This is a general problem of any deposition technique in this situation.

An alternative technique is of course the sputter deposition of U-Mo into a machined pocket (as previously discussed and demonstrated with Cu in figure 4.14b), and the sealing of this structure by the sputter deposition of a cladding cover layer. This method however suffers from the previously discussed general disadvantages when a U-Mo fuel kernel should be sputter deposited. A mixed solution, that means a conventionally produced U-Mo kernel that is placed inside a cladding plate with machined pocket and sealed by sputter deposition of a cover layer from cladding material, is also inefficient, as it would require an additional processing step to bond the U-Mo kernel to the cladding and thus provide no benefit to the currently used technique.

Pre-cladding A possible use for a sputter deposited cladding could be the 'pre-cladding' concept. A thin layer of the actual cladding material with several micrometers thickness is applied to a U-Mo fuel foil by sputter deposition, which takes only a few hours. The fuel foil is then fully cladded by a so-called 'pre-cladding', which has an excellent adhesion to the fuel foil and no step profile. The pre-cladded fuel foil is then laminated into a massive cladding plate with machined pocket and sandwiched with a cladding cover plate. This sandwich can easily be roll bonded afterwards, as the bonding process itself has only to bond cladding material with cladding material.

The pre-cladding concept could thus avoid the disadvantages of solely sputter deposited cladding but conserve its advantages.

Application I: Fuel fabrication

number and description	size [mm x mm]	coated with	cladded with
2 U-10Mo minifoils	28 x 30	Zry-4	AlFeNi
6 U-10Mo minifoils	60 x 30	Zry-4	AlFeNi
1 U-10Mo full size foil	600 x 60	Zry-4	AlFeNi
1 U-10Mo full size foil	600 x 60	Ti	Al

Table 4.4: U-10Mo minifoils and full size foils that have been coated so far by sputtering.

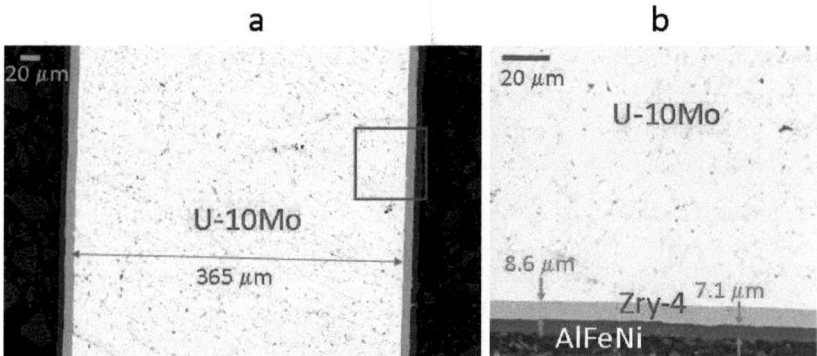

Figure 4.20: (a) Cross section of a U-10Mo foil sample sputter coated with Zry-4 as diffusion barrier and AlFeNi as pre-cladding. (b) The diffusion barrier layer has a thickness of approx. 9 μm, the pre-cladding layer approx. 7 μm.

U-10Mo foil pre-cladding The pre-cladding concept seemed more promising to us as the complete sputter cladding concept. Thus we decided to apply only pre-cladding layers during our U-Mo foil coating experiments. We used the U-10Mo foils listed in table 4.3, that had already been coated with diffusion preventive materials, and applied pre-cladding layers of Al and AlFeNi to them. Table 4.4 lists, which pre-cladding was applied to each sample.

The applied pre-cladding thickness was between 5 - 10 μm for all U-10Mo foils and minifoils. Figure 4.20 shows a cross section of a foil coated in this way.

Conclusion We demonstrated, that small and full size U-Mo fuel foils can easily be cladded with the materials Al and AlFeNi in a thickness of 1 - 100 μm. The deposition of cladding layers in the thickness required for fuel plates is however quite time- and energy-intensive: several ten hours of sputtering would be necessary at plasma powers in the kW range. Sputter deposition will to our opinion thus never be able to compete with the material throughput provided by the picture-frame cladding technique. The use of sputter deposition is however

4 APPLICATION

promising for surface preparation prior to the picture-frame cladding application.

4.2.4 Prospect

The fabrication process for fuel plates containing monolithic U-Mo fuel can be divided into three stages:

- **foil fabrication**, which encloses the alloying and casting of the U-Mo, rolling of the cast ingots to foils and shearing the foils to proper size
- **foil preparation**, which encloses the cleaning and coating of the fabricated U-Mo foils
- **plate fabrication**, which encloses the application of cladding to the coated fuel foils and the thickness finish

Techniques to realize each of these stages have been developed with great effort in the last years. These techniques currently allow a small scale production of monolithic fuel plates but are still in development. A first industrial scale production of fuel plates and fully assembled elements is supposed to be realized in the US within the next years [tec08].

The efficiencies for the different fabrication stages can be measured in terms of a yield factor, that denotes the material utilization in a certain process step. For the three stages listed above, the yield factors were estimated in [tec08] and are given as follows:

- foil fabrication: 76 %
- foil preparation: 85 %
- plate fabrication: 90 %

The total process yield is the product of these yields and thus estimated as 58 %, meaning that 58 % of the U-Mo allocated for fuel plate fabrication will actually end up as fuel plates while 42 % will be lost as scrap, chemical waste, casting losses, etc. Sputter processing can in principle be applied in all three stages of this fabrication process, but with varying effectiveness:

In the first stage, foil fabrication, sputter deposition could be used to replace the extensive ingot casting, ingot rolling and foil shearing procedures of the U-Mo material. Sputter deposition requires however the additional effort of fabricating the sputtering targets, which will also be produced by ingot casting and ingot milling[13]. Effectively the sputter processing thus only avoids the rolling and

---

[13]The presently used U-8Mo sputtering targets were fabricated in this way

Application I: Fuel fabrication

shearing step, as the casting step is necessary for both methods. Considering the times that are necessary for ingot rolling and foil shearing (approx. 8 h per foil, see [moo08]) on the one side and for the sputter deposition of monolithic foils (several ten to over one hundred h per foil, see section 4.2.1) on the other side, sputter processing is clearly not competitive to the conventional techniques. Considering the material utilization of rolling and shearing (84 %, see [tec08]) on the one side and for sputter deposition (4 - 9 %, see section 4.2.1) on the other side, sputter processing is again clearly not competitive. To our opinion it is therefore questionable, if sputter processing can ever be competitive to the currently used techniques for U-Mo foil fabrication, even if the material utilization in the sputter deposition process would be massively improved by the techniques shown in section 5.4.

In the second stage, foil preparation, sputter erosion could be used to assist or replace the cleaning process of U-Mo foils. The times for mechanical and chemical cleaning of U-Mo foils are given as several minutes per pass and repetition, until a sufficient surface quality is reached [moo08]. Sputter erosion cleaning requires several hours if it is used as the only cleaning method in the fabrication process, but also only several minutes if it is used in combination with either mechanical or chemical cleaning (see section 4.1.2). A comparison of the quality of surface cleanliness is unfortunately not possible for chemical cleaning and sputter cleaning, as the microscopic surface structure of the chemically cleaned surface is not known. A comparison of mechanically cleaned and sputter cleaned surface (see figure 4.10) indicates however, that sputter cleaned surfaces always show a higher degree of cleanliness compared to the conventional techniques. A combination of the currently used cleaning techniques as pre-cleaning steps and the sputter erosion cleaning technique as cleaning finish thus seems us to be a beneficial advancement of the foil preparation process.

The most promising use of sputter processing in the second stage could however be foil coating by sputter deposition. The major yield losses during foil preparation are attributed to foil rejections due to unacceptable coating results of the currently used coating techniques [tec08]. Our experiments with the sputter coating of previously sputter cleaned monolithic U-Mo foils however clearly show, that sputter processing allows to standardly produce a coating quality that is superior to the standard quality of the conventional coating techniques co-rolling and thermal spraying (compare section 4.2.2 to [moo08]). The times required to apply these coatings (several hours for coating thicknesses of 1 - 25 μm, see section 4.2.2) is however relatively long compared to the conventional techniques (co-rolling requires no time, as it is part of the foil fabrication, thermal spraying takes several minutes, see [moo08]). Sputter processing could thus reduce the number of unacceptable coating results significantly, but would increase the duration of the second stage by several hours. We estimate, that the yield factor of the second

4 APPLICATION

fabrication stage, that is currently estimated as 85 %, could be increased to at least 90 - 95 % by the use of sputter processing, which would rise the total process yield to 62 - 65 % and justify the increase of processing time. To our opinion, the foil coating by sputter deposition could therefore also be a beneficial advancement of the foil preparation process.

In the third stage, plate fabrication, sputter processing could be used for cladding application and replace the two currently investigated bonding processes HIP and FB. The yield factor of the conventional process is estimated with 90 % and results from plate rejections due to incomplete bonding of fuel foil to cladding. Sputter deposited cladding is supposed to show a smaller percentage of incompletely bonded plates and could allow to reach a yield factor of 90 - 95 %, but would require several ten to over one hundred hours of fabrication time per foil (see section 4.2.1). Compared to the necessary processing times of the conventional techniques (HIP and FB each require several hours per plate, see [tec08]), sputter processing would thus increase processing times significantly but the yield factor only marginally. To our opinion it is therefore questionable, if sputter processing is competitive to the conventional techniques in the phase of plate fabrication.

In summary, the application sputter processing in fuel plate fabrication seems reasonable only in the stage of foil preparation.

4.3 Application II: Scientific samples

In the process of developing a U-Mo based high density nuclear fuel it is mandatory, to study and understand the properties and behavior U-Mo alloys. All accordant experimental investigations require however, that appropriate U-Mo samples are available to the experimenters. We investigated sputter deposition in this context, as it allows to produce scientific samples that can be tailored to individual experimental requirements.

We produced sputter deposited samples for two different types of experiments:

- for irradiation experiments, that allow to study IDL formation

- for thermal diffusion experiments, that allow to study solid state reactions in U-Mo

The irradiation experiments were conducted by our group, the thermal diffusion tests were performed by Leenaers from SCK/CEN.

4.3.1 Irradiation experiments

A major activity in the fuel conversion research program at FRM II is the investigation of irradiation induced reactions in the interface regions of fuel and cladding. The interface regions are of particular interest, as the IDL formation during irradiation starts here. IDL have proven to be the key for phenomena like breakaway swelling, delamination or the appearance of hot spots (see chapter 1.2.4). Therefore it is essential to understand the occurrence, growth behavior and the properties of IDL to control them and to avoid those undesired phenomena during reactor operation.

Heavy ion irradiation The classical way to investigate radiation damage and radiation induced reactions is to irradiate test samples and test plates with the material combinations of interest inside a reactor core ('in-pile') under conditions that are as close as possible to the original fuel conditions simulating the complete nuclear burnup. After in-pile irradiation, the radioactivity of such test samples is in the range of 10^{11}Bq/gU [cle05], which prevents their near-term examination and makes a storage for months or years necessary to let the gross of activity wear off. The examination in hot cell facilities can start after that.
As this classical way of investigating radiation damage is quite time consuming and cost intensive, a quicker and cheaper alternative was looked for. Irradiation with heavy ion beams from particle accelerators is long known as an effective technique to simulate the structural modifications of in-pile irradiation. Wieschalla [wie06],[pal06],[jun10] could show, that an IDL quite similar to those produced in-pile in dispersed U-Mo/Al can also be produced 'out of pile', if dispersed fuel samples are irradiated with heavy ions that are in type and energy similar to typical nuclear fission products. This observation is not surprising, as the IDL produced by in-pile irradiation is mainly caused by the fast and heavy fission fragments generated in the fuel during fission, which are nothing else than fast heavy ions in matter.

MAFIA Our group uses the 'Munich Advanced Fuel Irradiation Apparatus' (or: 'MAFIA') that is shown in figure 4.21 for heavy ion irradiation (or: 'HII') [jun11]. MAFIA allows the defined HII of small material samples to simulate the radiation damages created by fission products. The advantage of this kind of irradiation is, that the samples will not be activated during HII and that several hours of irradiation time are sufficient to simulate the damages of several months of in-pile time.

By examination of irradiated samples via surface electron microscopy (SEM), x-

4 APPLICATION

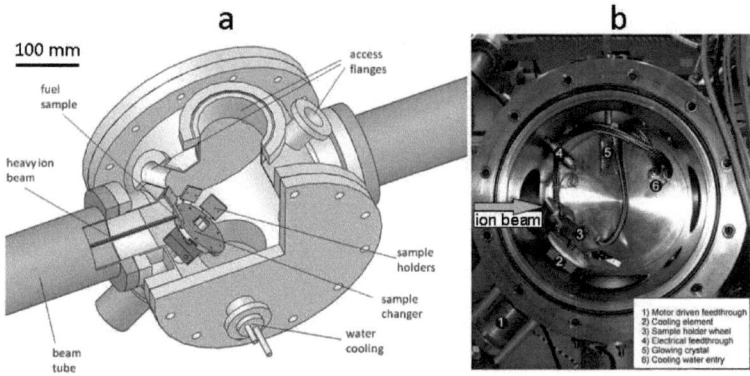

Figure 4.21: (a) Cut of the MAFIA setup for heavy ion irradiation experiments. The sample chamber is attached to the +10° beam line at the tandem accelerator of the Maier-Leibnitz Laboratory (or: MLL) in Garching. If the beam line is active, the heavy ion beam crosses the chamber along its middle axis. Several irradiation samples can be mounted into the sample changer, that automatically places them in the focus of the beam. Only one sample is irradiated at a time, and if the irradiation is finished the sample changer automatically switches the next one into the beam. The sample changer is temperature controlled, that means the irradiation samples can be irradiated at defined temperatures. The ion flux reaching the sample is monitored via a Faraday-cup, that is inserted into the beam at regular intervals. (b) View into the MAFIA sample chamber (picture from [jun11]).

ray diffraction (XRD) and optical microscopy (OM) before and after irradiation it is possible to study the modifications and damages induced in the sample material.

The first irradiation samples for MAFIA were provided by courtesy of AREVA-CERCA. A continuous supply with fuel samples on demand could however not be provided by AREVA-CERCA, as every sample had to be produced in Romans (France) and to be shipped to the FRM II site, which is a time consuming and expensive procedure. The advantage of MAFIA, which is the possibility of a quick and easy sample irradiation and examination, can however only be fully utilized, if an adequate supply of irradiation samples is secured. This source of supply should be able to provide samples at least as quickly as they can be irradiated and as many as can be irradiated. It should further allow to be flexible in the exact sample composition and ideally be on site to avoid cumbersome sample shipment. It was thus nearby to test, whether sputter deposited samples fabricated in our tabletop reactor could as well be used for HII.

Deposited samples Three types of sputter deposited samples could be of relevance in HII (see figure 4.22). First, a free standing monolayer system could allow to investigate the change of properties in a material due to irradiation damages. The thermal conductivity of U-Mo and its change during irradiation could possibly be measured in this way. Second, a two layer system should illustrate the irradiation behavior at the interface of two materials in contact with each other, especially the IDL formation in the fuel/cladding system. This might open the possibility to systematically test the compatibility of different materials with each other at in-pile conditions. Third, a three layer system might be used to test and develop barrier layers between fuel and cladding that reduce or avoid IDL formation. HII could allow to conduct a systematic screening procedure to determine the most efficient barrier material and an optimum barrier thickness.

The thickness of the sputter deposited layers for HII is determined by the average range λ of the heavy ion beam in matter, and thus by the ion kinetic energy. As the ions are used to simulate fission products, typical ion energies will be of the order of 80 MeV[14], which corresponds to an average range λ of the order of several micrometers for all solid materials. In the case that the ion beam should completely be stopped inside the layer, the layer thickness has to be greater than the average range of ions. In the case that radiation induced interactions between different layers are of interest, the layer thickness has to be small enough, so that

---

[14] A single fission reaction of ^{235}U releases about 200 MeV. From this energy a fraction of roughly 160 MeV is distributed to the heavy fission products, that have, dependent on their mass, an average energy of around 80 MeV each.

4 APPLICATION

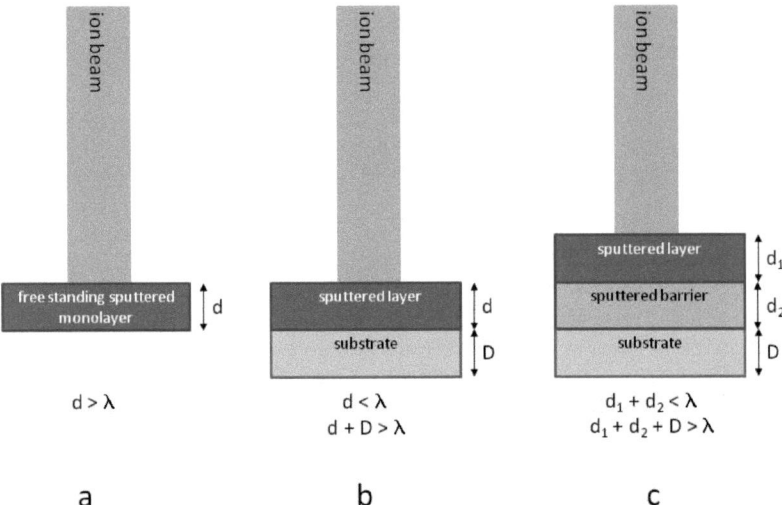

Figure 4.22: Sputtered geometries for HII samples: (a) Free standing monolayer to investigate a change of material properties during irradiation. The thickness d of the layer has to be larger than the range λ of the heavy ions to stop all ions inside the material. (b) System with one layer sputtered on a substrate to investigate the interaction between two materials during irradiation. The thickness d of the sputtered layer has to be smaller than λ, so that the heavy ions can actually reach the substrate and enable an interaction. The thicknesses of sputtered layer d and substrate D have to be larger than λ to stop all ions. (c) System with two different layers sputtered on a substrate to test the quality of barrier structures. The thicknesses d_1 and d_2 of the sputtered layers have to be smaller than λ so that the heavy ions can reach the substrate. The thicknesses d_1, d_2 and D have to be larger than λ to stop all ions.

Application II: Scientific samples

material	mean ion range [in μm]
U-8Mo	5.01
Ti	8.64
Zr	8.17
Zry-4	8.18
Bi	8.84
Al	12.8

Table 4.5: Ranges of ^{127}I ions with an energy of 80 MeV in different materials (calculated by SRIM). The calculated values allow to estimate, how thick the layers in one, two and three layer systems have to be.

most ions can reach both layers.

In two or three layer systems the order of layers is given by the trajectory of the fission products. In actual fuel plates, the ions respectively fission products will be emitted from the fuel and be stopped inside the cladding. Therefore the sample top layer facing the ion beam should be U-Mo and the bottom layer (respectively substrate) should be a cladding material.

The necessary coating thicknesses for one, two and three layer systems can be determined by simulations with the program SRIM (see appendix A1). The samples in MAFIA are usually irradiated with the isotope ^{127}I at an energy of about 80 MeV (see subsequent paragraph). With SRIM it is possible to calculate the mean ion ranges in the desired layer systems for this energy and ion type (see for example table 4.5), and thus to estimate the optimum layer thicknesses. The coatings can then be applied to the substrate surfaces in the tabletop sputtering reactor.

Sample production For our first test of sputter deposited samples for HII we decided to irradiate two layer systems of U-Mo and Al, as they are well comparable to the already known dispersed U-Mo/Al systems. Layers from U-Mo provide only a relatively small ion range of about 5 μm (see table 4.5). To allow the ions to pass the U-Mo top layer and to reach the Al bottom layer, the thickness of the U-8Mo had therefore to be smaller than 5 μm. We decided to use a U-8Mo thickness of 2 - 3 μm for our samples.

The heavy ion beam spot in MAFIA has dimensions of about 2 mm x 6 mm, which is therefore the minimum required size for an irradiation sample to utilize the beam completely. It turned out during our experiments, that a size of 10 mm x 10 mm is convenient for sample handling. We therefore used pieces of pure Al (99.9995% purity) respectively Al-6061 in these size with a thickness of approximately 200 μm as sputtering substrates for sample production. They were highly

4 APPLICATION

polished on one side on a polishing disc with a diamond polishing liquid (diamond grain size 1 μm) to create a clean and flat surface with a roughness of ≤ 1 μm.
The Al substrates were coated in the tabletop reactor with U-8Mo, as it was the only U-Mo alloy available as sputtering target. The processing parameters were kept constant for all samples to have comparable results. The plasma power was chosen to be 300 W. According to the U-I-characteristics, the plasma voltage at this power was about 300 V, the plasma current was about 1 A. The sample table temperature was kept at about 15 °C. The Ar pressure was $3 \cdot 10^{-3}$ mbar. According to the Thornton zone model, the growing layers should all be of T structure.

Irradiation We conducted a first HII at the tandem accelerator of the Maier-Leibnitz Laboratory (or: MLL) in Garching [jur10]. We primarily wanted to test, if the deposited samples would survive HII, and if an IDL would actually form. We were also interested in basic questions like, whether the substrate topography would have an influence on IDL formation.
The choice of the heavy ion type used for HII was a compromise between the necessity of a typical fission fragment isotope and technical feasibility. The two maxima of the statistical fission fragment distribution curve for U fission are located between the atomic mass numbers 90 to 100 and between 135 to 145 [sta07], therefore heavy ions with a mass in this range would be preferable for HII from the physical point of view. Unfortunately this mass ranges cannot be covered completely by the available types of ions at the tandem accelerator[15]. In the lower mass range between 90 to 100 only ion beams of the elements Zr, Nb, Mo and Ru can be provided, in the higher mass region between 135 to 145 no element is available at all [pri10]. The closest available isotope to the higher mass region is ^{127}I. From simulations with SRIM we knew, that heavy fission fragments respectively heavy ions from the mass range around 135 to 145 cause much more irradiation damage as the light ones between 90 to 100, as they are stopped much faster and thus deposit their energy in a much smaller volume. Additionally Wieschalla successfully used ^{127}I for fuel irradiation already in 2006 [wie06],[pal06],[jun10]. Therefore we decided to irradiate our samples with ^{127}I as well, although this isotope is not a very frequent fission product.
We chose the ion energy to be 80 MeV, as this is a typical fission product energy. The tandem accelerator can provide a ^{127}I flux of up to $5 \cdot 10^{11} \frac{ions}{s \cdot mm^2}$ at this energy with a charge state of +6. We decided to irradiate the samples under normal incidence, to have no angular dependency in the irradiated area. The samples were irradiated until the affected volume had reached a mean ion density

---

[15]Primarily the electrochemical behavior, but also other aspects determine, if an element can be ionized and accelerated in the tandem accelerator and hence if it is available for HII.

Application II: Scientific samples

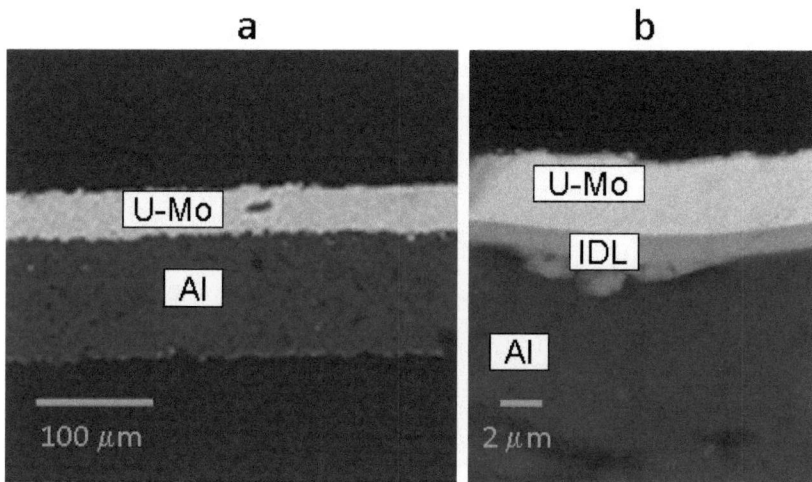

Figure 4.23: (a) SEM-BSD picture of a sputtered two-layer system with U-8Mo on Al (from [jur10]) (b) SEM-BSD picture of a U-8Mo/Al sample irradiated at 200°C (from [jur10]). The IDL that formed at the interface between U-Mo and Al is clearly visible. The Al substrate was not polished and the IDL was thus clearly thicker as in the case when the Al substrate was polished.

corresponding to the fission fragment density in a fuel with $\geq$ 50% nuclear burnup[16]. The affected volume in this case is defined as the HII beam spot size (2 mm x 6 mm) times the mean ^{127}I ion range. The irradiation time to reach this value generally was around eight to eleven hours for one sample. The needed period of time varies for each sample due to the usually quite instable fluence of the ion beam and due to the different layer thicknesses in the samples. It has to be adjusted during each single sample irradiation.

Examination The irradiated samples were embedded into resin and cut into half. The cutting edge was polished and examined by SEM. In two of the samples an IDL could be observed (see figure 4.23b).

Table 4.6 lists the samples that were irradiated by us.

Apparently sputter deposited U-Mo/Al systems show the same IDL formation during HII as dispersed U-Mo/Al samples. A comparison of samples 1 and 2, both with polished substrates, that were irradiated at 100°C respectively 200°C

---

[16]It should be noted, that the maximum reachable nuclear burnup for a future FRM II fuel will be around 14 % [bre11]. By irradiating to $\geq$ 50% one tries to study the behavior of the fuel far beyond this value to have a safety margin.

4 APPLICATION

sample number	substrate surface	U-8Mo layer [μm]	temperature [°C]	IDL [μm]
1	polished	2.3	100	0
2	polished	2.3	200	0.6
3	unpolished	3.4	200	1.0

Table 4.6: Three sputter deposited U-8Mo/Al samples that were irradiated by us. For all three samples the ion fluence during HII was $7.01 \cdot 10^{16} \frac{ions}{cm^2}$ and the reached ion density was $8.54 \cdot 10^{21} \frac{ions}{cm^3}$, which corresponds to an equivalent burnup (EBU) of 67%. The temperature was measured with a PT-100 sensor.

gives evidence, that the process of IDL formation is not only dependent on irradiation conditions but also on temperature. A threshold temperature value between 100°C and 200°C might exist. A comparison of sample 2 and sample 3, one with polished and the other one with unpolished substrate, gives evidence, that the substrate topography affects IDL formation. Rougher substrate topography means increased IDL formation. This behavior could be expected, as the substrate topography directly determines the interface area between the different materials.

Conclusion We could show, that sputter deposited samples are suited for HII, and that IDL formation can clearly be observed. We also found clear evidence for the expected influence of substrate topography and temperature on IDL formation.

4.3.2 Thermal diffusion experiments

An effect of major importance in U-Mo fuels is the IDL formation during in-pile conditions. As mentioned, IDL formation is the result of a radiation enhanced thermal atomic migration effect and caused by a solid state reaction of U and Al. The application of an appropriate third material like Si in between U-Mo fuel and Al cladding could prevent IDL formation, if the U gets immobilized by forming a compound with the Si. A method to study such solid state reactions between U-Mo and other materials are thermal diffusion experiments.

Diffusion experiments The efficiency of Si addition for IDL prevention can be estimated from the reaction kinetics of Si. Generally, the kinetics of a thermally activated chemical process can be described by the well-known Arrhenius equation:

$$k = A \cdot exp^{-E_a/k_B T}$$

Application II: Scientific samples

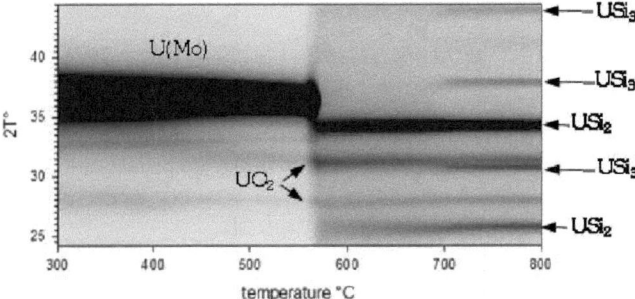

Figure 4.24: Figures from [lee10]: In-situ XRD results for ramp anneal of a Si substrate covered with a 500 nm U-Mo layer and a 30 nm Si layer (both produced by sputter deposition) at a rate of 0.2 °C/s. The transit between U-Mo and USi$_X$ at about 570 °C is clearly visible.

with k being the growth rate coefficient, A the reaction coefficient, E_a the activation energy and T the temperature. The value of E_a in this equation is an inherent characteristic of a particular chemical reaction. From the knowledge of E_a for different Si reactions it is thus possible to estimate the Si reaction kinetics at a certain temperature T.

For a solid state reaction E_a is usually determined by measuring the time that is needed at a fixed temperature to initiate the reaction. As Leenears states, an alternative method is the use of ramp annealing to determine the formation temperature of a reaction product [lee10]. In this method, the samples are heated with different fixed heating rates and the sample crystal structure is monitored online by X-ray diffraction. Figure 4.24 shows exemplary the XRD results for a ramp annealing of a sample consisting of a 500 nm U-8Mo layer with a Si substrate at a heating rate of 0.2 °C/s. It is clearly visible, that at approximately 560°C the U reacts with the Si forming USi$_2$ and USi$_3$. The ramp heating rate determines the integrated thermal budget to which the sample has been exposed prior to reaching a certain temperature. The integrated thermal budget will be much higher for small heating rates than for large ones, as the sample will spend longer time at each temperature. As a result, the value of formation temperature will depend on the ramp heating rate.

According to [lee10], it is possible to determine E_a by using the so-called Kissinger equation:

$$ln\left(\frac{\frac{d}{dt}T}{T_f}\right) = -\frac{E_a}{k_B T_f} + const.$$

and by determining the formation temperature T_f from the maximum peak intensity of the XRD measurement.

131

4 APPLICATION

Study of the U/Si solid state reaction We prepared samples consisting of U-8Mo sputter deposited on Si wafers. The wafers were first coated with a layer of 0.1 - 1 µm U-8Mo and then covered with several ten nanometers Si respectively Zr coating for oxidation prevention. Leenears from the Belgian Nuclear Research Center (SCK/CEN) performed ramp annealing and XRD measurements on these samples to investigate the solid state reaction between U and Si [lee10]. She used different ramp heating rates from 0.2 - 3.0 °C/s, and recorded the XRD plots shown in figure 4.25.

The growth of an USi_2 and an USi_3 phase could be observed for all prepared samples. Complementary experiments of a Si layer deposited onto a U-Mo substrate led to the formation of U_3Si and U_3Si_2. Leenears used Kissinger analysis on the XRD plots to determine the activation energy of the silicide formation. She determined E_a values of 3.5±0.5 eV for USi_2 and 4.4±0.6 eV for USi_3, respectively 3.1±0.3 eV for U_3Si_2 and 5.1±0.8 eV for U_3Si.

Conclusion The most important solid state reactions between U-Mo and Si could be examined by thermal diffusion experiments with sputter deposited samples, and the according activation energies E_a could be determined. These factors are necessary to estimate the reaction kinetics of Si and U-Mo at a given temperature. This kind of measurement is only possible, because the sputter deposition process allows the application of thin layers ≤ 1 µm of different materials, that can be penetrated by XRD measurement. Other relevant solid state reactions concerning U-Mo could be studied in the same manner.

4.3.3 Prospect

The combination of sample production in the tabletop sputtering reactor and the HII of these samples in MAFIA is from our experience an effective way to quickly emulate the in-pile behavior of material systems containing U-Mo.

As mentioned, sputter deposition allows to form very plain and homogeneous layers of arbitrary materials on arbitrary substrates. Conductive elements can be processed by DC or RF sputtering, non-conductive elements by RF sputtering. Alloys can be prepared by co-sputtering[17] of pure elements, compounds by reactive sputtering. Therefore systems of all possible combinations of U-Mo and cladding alloys, barrier materials and additives can be realized easily. MAFIA on the other side allows to produce irradiation damages in these samples that

---

[17]Co-sputtering denotes the parallel use of several different sputtering targets with different materials and one substrate. It allows to mix the different ejected atom fluxes to deposit a film of mixed material.

Application II: Scientific samples

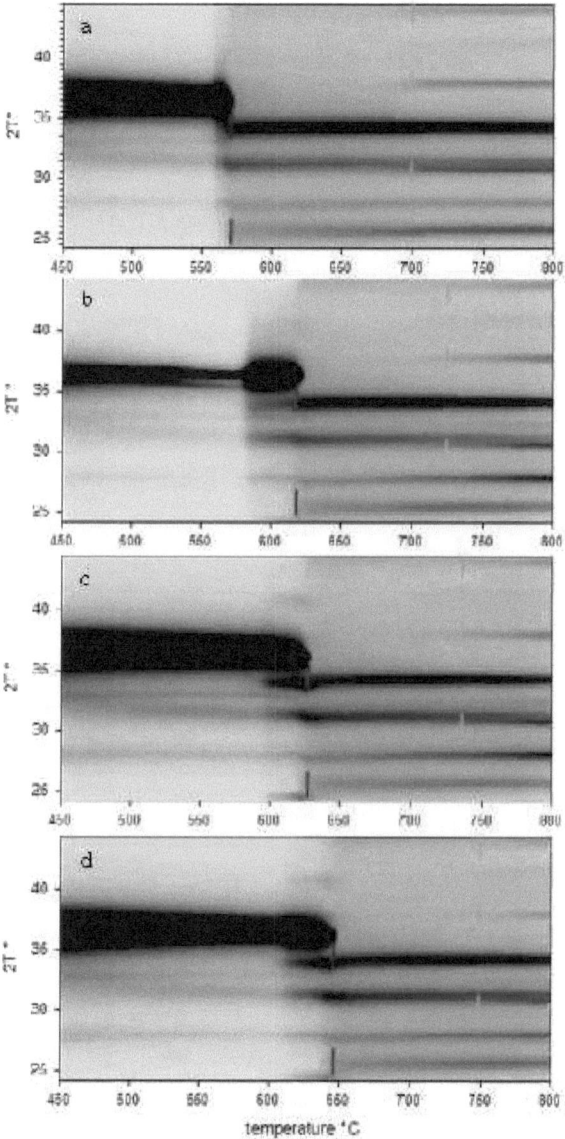

Figure 4.25: Figures from [lee10]: In-situ XRD results for ramp anneal of a Si substrate covered with a 500 nm U-Mo layer and a 30 nm Si layer (both produced by sputter deposition) at a rate of (a) 0.2 °C/s (b) 0.5 °C/s (c) 1.0 °C/s (d) 3.0 °C/s. The dark grey markers indicate the growth of a USi_2 phase while the light grey markers point out the formation of a USi_3 phase.

4 APPLICATION

are similar to those induced by nuclear fission and thus to simulate the effect of nuclear burnup for various temperatures and heat fluxes. Three types of experiments based on HII of sputtered samples seem promising:

- screening of additive materials
- optimization of diffusion barriers
- thermal conductivity measurement

Co-sputtering allows to produce U-Mo+X/Al+Y systems with defined additives X / Y both in U-Mo and Al. HII of such samples opens the possibility to systematically screen additives according to material and concentration in order to determine the most effective material in the most effective concentration to prevent IDL formation. In a similar way can multilayer systems of the structure U-Mo/X/Al with a barrier material X be produced either by DC or RF sputtering. For samples of this type, HII allows to screen the IDL preventive effect of layers from different materials X in different thicknesses and also the determination of the most efficient barrier material and an optimum barrier thickness [sch11].

A question of some interest that has hardly been investigated up to now is the thermal conductivity of U-Mo fuel and its change during reactor operation. The change of U-Mo thermal conductivity during reactor operation is expected to result from damages in the fuel material created by the fission reactions respectively by the resulting fission products. HII generates damages in U-Mo that are similar to those generated by in-pile irradiation. We thus expect, that the change in U-Mo thermal conductivity induced by HII could also be similar to the the change induced by in-pile irradiation. HII could thus allow to measure the evolution of U-Mo thermal conductivity without the disadvantage of activating the fuel samples. Even an online measurement seems to be easily feasible with HII.

Hengstler tried to show the feasibility of the mentioned concept for cast monolithic U-8Mo fuel samples [hen08]. Her samples had a thickness of 200 μm, the irradiated surface layer had however a thickness of only 5 μm, and the change of heat conductivity in the irradiated zone was apparently too small to resolve the effect. Hengstler identified the too large sample thickness, the too small sample size and the small number of available samples as the major problems of her measurement and suggested to use thin monolithic foil samples of maximum 5 - 10 μm thickness and a minimum size of 10mm x 10mm. As samples of that kind could not be produced conventionally Hengstler was not able to pursue her measurements.

By sputtering it is now possible to produce self-supporting monolithic U-Mo foil of the kind Hengstler suggests. A measurement of the HII induced change in thermal conductivity thus seems feasible.

Chapter 5

Advancement

This last chapter describes concepts for a quality improvement of the deposited films as well as for an advancement of the two sputtering reactors.

The film quality can strongly be improved by a reduction of reactor inherent gradients and by avoiding sample pollution during deposition. The fields of improvement for the reactors cover the extension of the material spectrum that can be deposited as well as the increase of target utilization.

5.1 Film gradients

Current status

During our experiments we observed the general appearance of thickness gradients in the deposited films, which means that the films have no homogenous thickness over the substrate surface (see figures 3.19 and 3.20). The thickness gradients in the two sputtering reactors result from the plasma geometry and position as well as from the cross-magnetron effect, which is also caused by plasma geometry. Moreover we have evidence, that the plasma geometry also causes a small compositional gradient, if multiatomic materials are deposited (see section 4.1.1). Both types of gradients are generally undesired, as they might impose problems for certain applications.

Concept for improvement

To our opinion it could be possible to reduce all kinds of gradients to an acceptable magnitude by simple methods. Three methods seemed reasonable to us:

- movement of plasma position
- controlled shadowing of the ejected material flux

5   ADVANCEMENT

- inactivation of target areas

Profile smoothing methods   As a first method we considered a movement of plasma position. The static geometry and the static position of the discharge plasma apparently have the biggest impact on the deposition profile and thus on the formation of gradients. It is thus nearby to use a dynamic plasma geometry or a variable plasma position to receive a blurring effect that can reduce gradients. The plasma geometry and position are both determined by the geometry and position of the magnetic assembly. A dynamic geometry of the magnetic assembly would thus lead to a dynamic plasma geometry. There are examples in literature, where a dynamic geometry of the magnetic assembly has been realized by a continuous rearrangement of the single elements of the magnetic assembly, for example in [fre87]. It is however also possible to completely replace the permanent magnetic assembly by a system of magnetic coils, that allow a complete control of the position, direction and strength of the magnetic field and thus of the plasma. We however considered all options for a dynamic plasma geometry as technically too extensive to realize, as a continuous motion of a static magnetic assembly and thus a variable plasma position seemed us much easier feasible.

As a second method we considered an installation of masks into the sputtering diode, that lead to a partial shadowing of defined substrate areas from the material flux coming from the target. The mask is effectively a massive material plate with a defined opening, that is located between target and substrate and prevents ejected material from certain solid angles to reach the substrate surface. The form and aperture of the opening will blind out certain areas of the target or directions of deposition and by that influence the deposition profile on the substrate[1].

The third method is denoted as 'inactivation' of target areas, which means that defined areas of the sputtering target can intentionally be excluded from ion bombardment. These inactivated areas thus will not face sputter erosion any more, and no material will be ejected from their surface towards the substrate. Inactivation is based on the effect, that for both DC and RF sputtering a negative voltage has to be applied to the sputtering target to generate an ion bombardment. If the voltage is however not applied to the whole target but only to certain parts of it, the ion bombardment will be limited to these particular areas. In a DC sputtering reactor this can easily be achieved, if parts of the electrically conductive sputtering target are replaced or covered by electrically non-conductive materials. If an electrical field is used to induce an ion bombardment on the

---

[1] It has to be mentioned, that the blinding out of directions of deposition will of course also influence the film growth process to a certain extent and thus influence film properties.

Film gradients

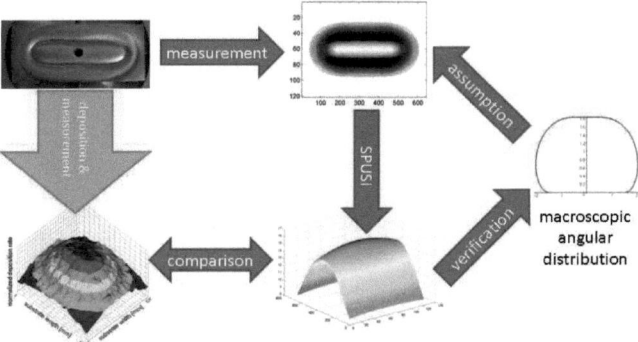

Figure 5.1: Principal approach to gain the macroscopic angular distribution of ejected material: first we assume a random but reasonable macroscopic ejection distribution. With SPUSI it is possible to simulate the deposition profile that would result from this particular distribution. The necessary pattern of active sputtering areas can be gained by measuring an actually used sputtering target. The deposition profile gained by simulation can now be compared to an actually measured deposition profile. If it matches, the assumed macroscopic ejection distribution is verified. Otherwise it has to be modified, and the procedure starts again.

surface of these non-conductors, the ion charge will accumulate and build up a reversed electrical field, that prevents further ion bombardment. A similar effect occurs in an RF sputtering reactor, if parts of the target are covered by grounded shields.

To our opinion one of these methods or a combination of them could reduce both thickness and compositional gradients effectively. As each of these measures requires however some effort to be realized in experiment, we decided to evaluate their individual effectivity by using a computer simulation.

Simulation of static configuration We tried to reproduce the experimentally measured deposition profiles of the tabletop reactor and the full size reactor with the program SPUSI as accurately as possible. SPUSI was developed by our group to simulate the sputtering process on a macroscopic scale. Its function is described in detail in the appendix A2.

We varied the macroscopic angular distribution via the matching coefficient x and calculated by that various deposition profiles. By a comparison between simulated and real deposition profiles, we were able to identify values of x, that sufficiently reproduce the measured profiles (see figure 5.1).

For the processes in both sputtering reactors we determined a value of $x \approx 5.5 \pm 0.1$ (see figure 5.2 respectively figure 5.3). The resolution r in this simula-

5 ADVANCEMENT

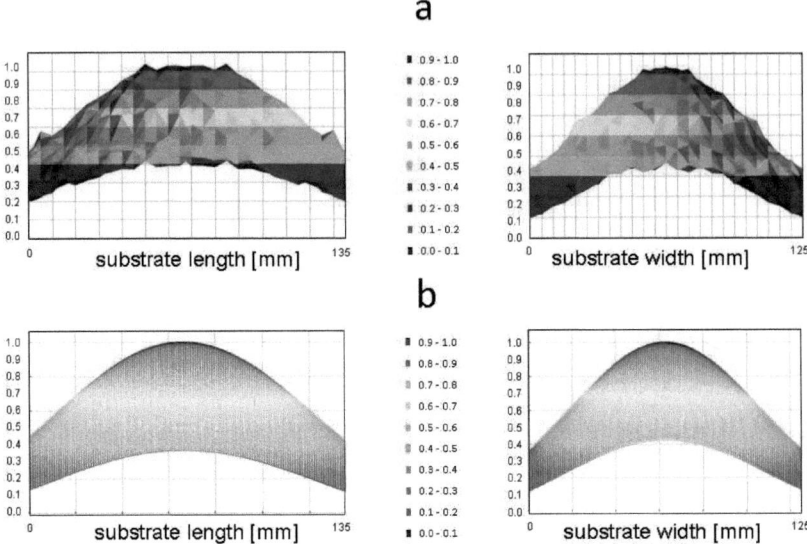

Figure 5.2: (a) Experimentally measured deposition profile in length (left) and width (right) of the tabletop reactor. (b) Simulated deposition profile in length (left) and width (right) of the tabletop reactor. The profile calculated by SPUSI is in good agreement to the measured profile, although SPUSI completely ignores the physical processes and just considers the geometry of the problem. Apparently the geometry is the main factor that is responsible for the macroscopic deposition profile of a sputtering target.

tion was 1 mm, the distance between target and substrate assumed to be 100 mm. As figure 5.2 shows, the deposition profile simulated by SPUSI is in good accordance to the measured one. This leads to the assumption, that the geometry and position of the discharge plasma is probably the most important factor that is responsible for the macroscopic deposition profile of a sputtering target, and that the microscopic physical ejection pattern only has a minor contribution. The simulated profile in figure 5.3 is still in accordance to the measurement, but the deviations are larger. The small contribution of the microscopic ejection processes apparently gets more pronounced in the larger geometry. SPUSI thus allows to simulate the macroscopic deposition characteristic in a static target/substrate configuration and the accordant gradients (see figure 5.4). The accuracy of the simulations is sufficient to use SPUSI as a tool to test the efficiency of measures to reduce gradients and smoothen the deposition profile.

Simulation of profile smoothing methods We tested the mentioned methods movement, shadowing and inactivation independently and in combination only

Film gradients

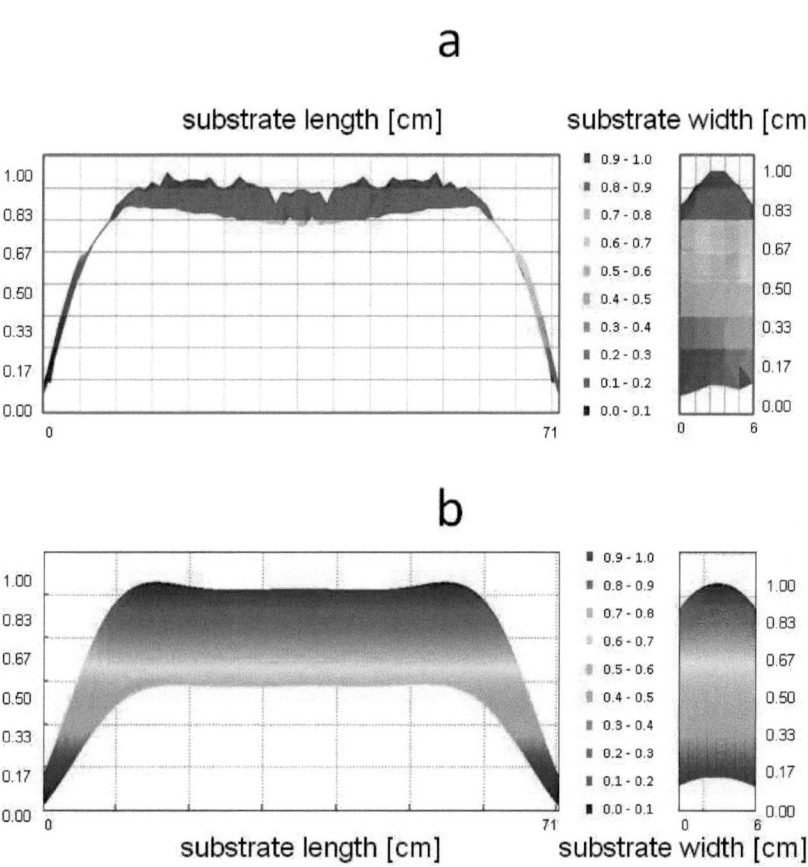

Figure 5.3: (a) Experimentally measured deposition profile in length (left) and width (right) of the full size reactor. (b) Simulated deposition profile in length (left) and width (right) of the full size reactor. The profile calculated by SPUSI is still in agreement to the measured profile in the central area, but deviations are bigger.

5 ADVANCEMENT

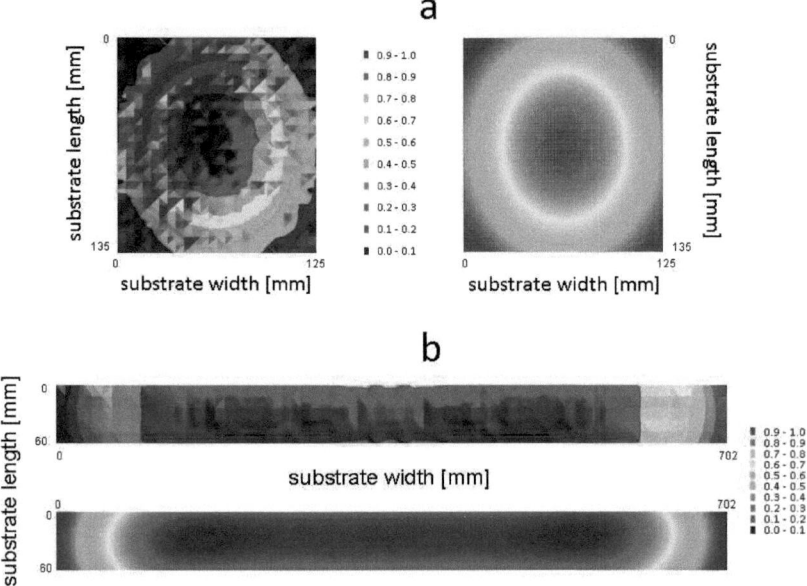

Figure 5.4: (a) Thickness gradient map for a deposit of size 125 mm x 135 mm produced in the tabletop reactor measured (left) and calculated by SPUSI (right). The calculated gradients are in good agreement to the measured ones. (b) Thickness gradient map for a deposit of size 702 mm x 60 mm produced in the full size reactor measured (top) and calculated by SPUSI (bottom). The calculated gradients still show a fair agreement to the measured ones.

Film gradients

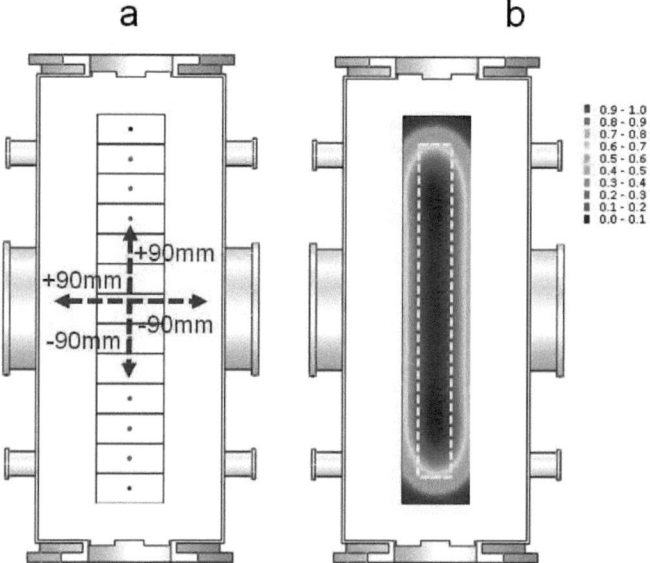

Figure 5.5: (a) Simplified cut view of the processing chamber in the full size sputtering reactor. The figure illustrates the position of the sputtering target and the space that is available around this target for installations and reconstruction. The full size reactor provides about 90 mm in each direction at the maximum. (b) Simplified cut view with static deposition profile beneath the sputtering target. The highlighted area sized 600 mm x 60 mm illustrates the projected position of the U-Mo fuel foil on the substrate carrier, where the thickness gradients should mainly be reduced.

for the geometry of the full size reactor. The vacuum chamber of the tabletop reactor was considered as too small to actually allow a realization of any of the methods.

Our aim was to reduce the thickness gradient in an area of 600 mm x 60 mm size (dimension of the available U-10Mo foils) in the centre of the substrate as far as possible (see figure 5.5b). The limiting factors were primarily the the inner dimensions of the processing chamber (see figure 5.5a).

We used the procedure illustrated in figure 5.6 to test the various procedures. In our simulations we observed, that relative motion of the elements target, plasma and substrate to each other in general produces a blurring effect as it was expected. The impact of this blurring on the evolution of gradients on the substrate is however strongly dependent on the particular path and speed of motion and can decrease but also increase the resulting gradients.

Mask structures generally seem to be an inappropriate mean to reduce overall thickness gradients in a sputtered deposit. They have the effect of blinding

5 ADVANCEMENT

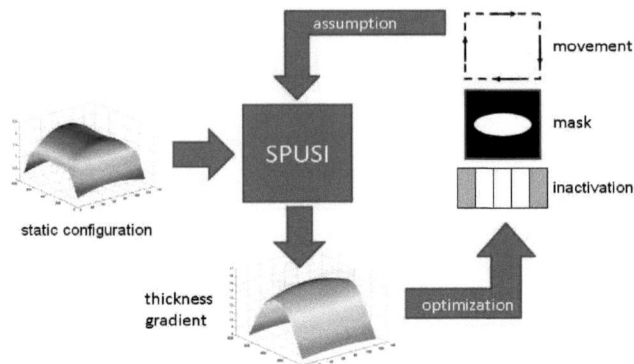

Figure 5.6: Principal approach to test the effectivity of profile smoothing methods. Prerequired parameter for SPUSI is the deposition profile of the static diode configuration. SPUSI allows to simulate deposition profiles if a smoothing method is applied to this configuration. From the homogeneity of the resulting profile one can estimate the effectivity of the method.

out the deposition in the masked regions. In unmasked regions, the deposition will however only be marginally disturbed, and will thus show nearly the same thickness gradients than before. A mask is therefore considered as an appropriate mean to completely blind out material flux or to reduce deposition to the walls of the sputtering reactor. It is however ineffective as a mean to smoothen the deposition profile. In a similar manner the inactivation of target areas also blinds out material flux, comparable to a mask structure that is mounted directly onto the target surface.

We came to the result, that from the three examined methods apparently only motion can efficiently avoid gradients without being too extensive in realization. Gradients along only one direction (length or width) can easily be smoothened by a relative movement along this particular direction. To receive an efficient smoothening effect, the local speed of motion has to be adapted to the local thickness gradient (see figure 5.7). We observed, that a speed profile shaped identically to the thickness profile can assure a nearly gradient free deposition profile in the substrate central area (see figure 5.8).

The thickness gradients in the static deposition profile are present both in length and in width. To smoothen this profile it is therefore necessary to superpose a motion along the length axis with a motion along the width axis. To get a sufficient homogeneity over the whole area, one of these motions should moreover be much faster than the other one (see figure 5.9). The resulting smoothened profile is shown in figure 5.10.

Film gradients

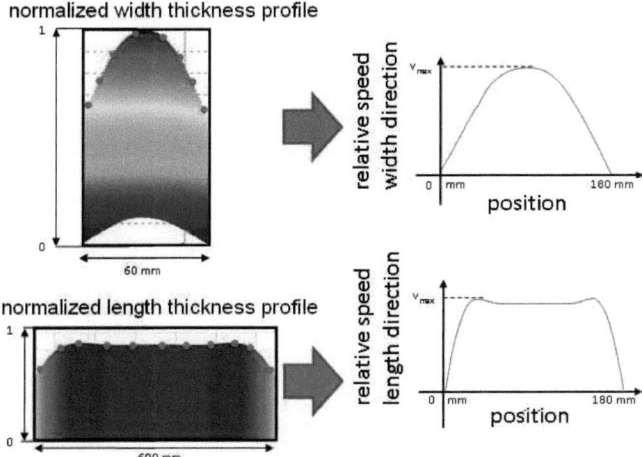

Figure 5.7: From the shape of the thickness gradients in width (top) and length (bottom) in the static deposition profile we obtained the shape of the speed profile for the relative motion in width and length. We identified speed profiles of these shapes to be the most efficient ones for thickness profile smoothing.

In the given geometry, that means with a target of 702 mm x 122 mm in size, a processing chamber with 90 mm free space in each direction and a substrate central region of 600 mm x 60 mm, we were able to reduce the width thickness gradient to a value of $\frac{minimumthickness}{maximumthickness} > 0.99$ with the shown method. The length thickness gradient could be reduced to $\frac{minimumthickness}{maximumthickness} > 0.65$ in the length of 600 mm and to > 0.99 in a length of 350 mm. A further reduction could not be reached in the given geometry, but would be feasible if the geometry of the setup could be changed.

Conclusion By simulations with the program SPUSI we identified a relative motion of either magnetic assembly, target or substrate towards the other components as most effective and simplest method to reduce thickness gradients in sputter deposited films. We gained the best results for a two dimensional motion with a variation in speed in each dimension. If the speed profiles and the relation of speed in length and width are chosen adequately, the thickness profile in the central area of the substrate can nearly be completely suppressed. The space limitation given by the size of the processing chamber makes it however impossible, to reach a nearly gradient free deposition zone of 600 mm x 60 mm in the full size reactor with the given setup geometry. Only a gradient free deposition zone of 350 mm x 60 mm could be reached.

5 ADVANCEMENT

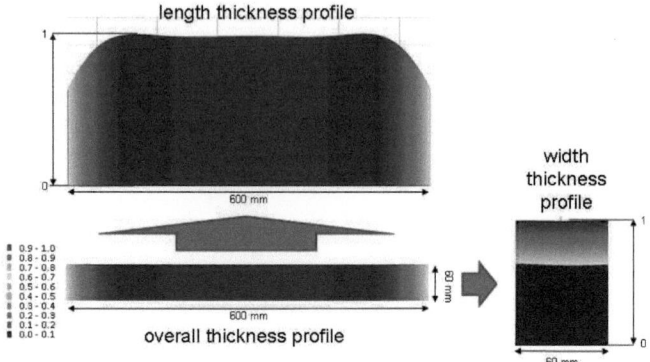

Figure 5.8: Thickness profile for a relative motion in width with a speed profile according to 5.7. The thickness gradient in width has nearly disappeared, the gradient in length is unchanged.

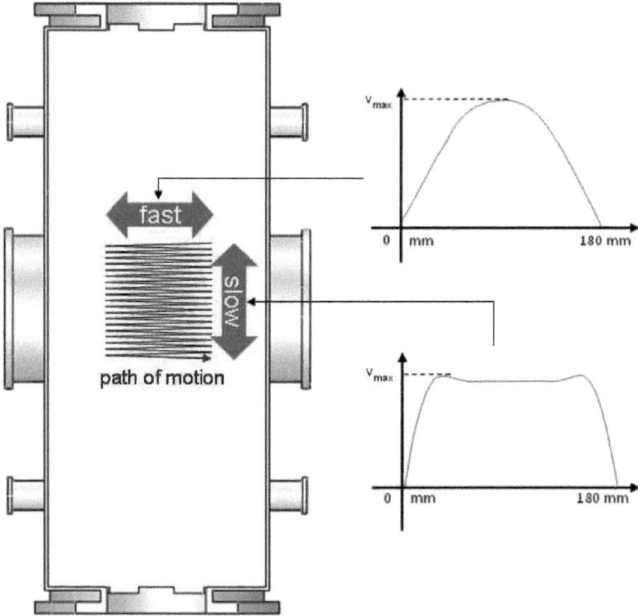

Figure 5.9: The speed profiles gained by the method shown in 5.7 can be used to smoothen a two dimensional thickness profile by superposition of a motion along the length axis with a motion along the width axis. To get a sufficient homogeneity over the whole area, one of these motions should moreover be much faster than the other one. The size of the processing chamber limits the dimension of the motion path.

Film pollution

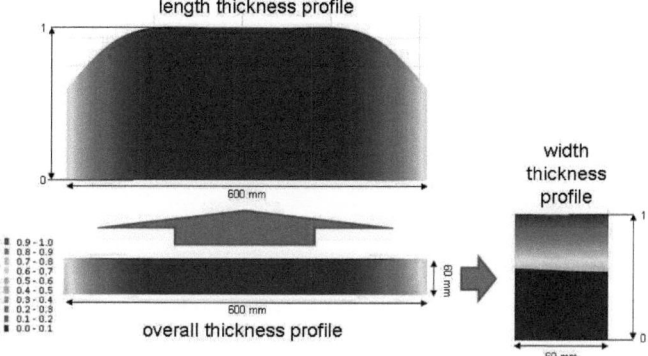

Figure 5.10: Thickness profile for a relative motion in width and length with speed profiles according to figure 5.7 and a motion path according to figure 5.9. The thickness gradients in width and length have nearly disappeared in the central region of the deposit. The central region is however much smaller than the 600 mm x 60 mm size we wanted to smoothen, as the size of the processing chamber does not allow more extended motion paths.

5.2 Film pollution

Current status

During film deposition we observed the occasional occurrence of pollution and growth defects in the sputtered deposits, even when the substrates had been prepared carefully before. Primarily films that were produced in the tabletop reactor were frequently subject to this kind of effect and showed flaking as well as macroscopic particles, that had been embedded into the sputtered film. As reason for the pollution and the flaking we determined macroscopic particles, that trickled onto the substrate surface during the deposition process (see figure 5.11). These particles polluted the samples already during manufacturing, and locally caused the growth of flake structures or nodular defects.

As origin of the particles we could identify the walls of the vacuum chamber. Material, that had been deposited to the walls of the sputtering reactor in present and previous deposition runs, accumulates over time and forms a thick porous layer covering the walls of the vessel. It turned out, that this layer is subject to increased oxidation due to air contact as long as the vacuum vessel is open. The vibration of the vacuum pumps during the sputtering process then leads to a steady release of small oxide and other material particles from the walls.

5 ADVANCEMENT

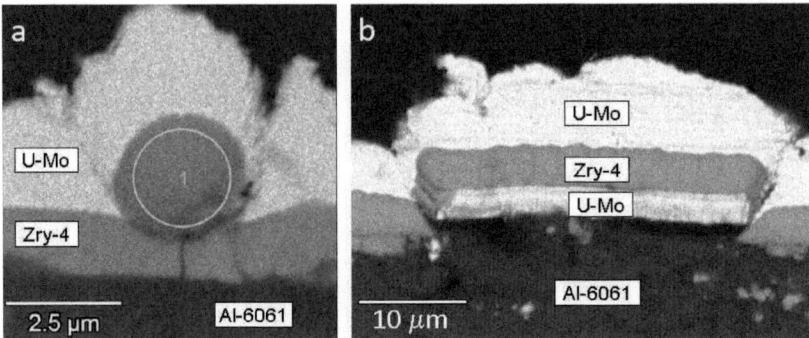

Figure 5.11: (a) Spherical Zry-4 particle, that apparently dropped onto the sputtered film after the deposition of the Zry-4 layer and prior to the deposition of the U-Mo layer. The particle was embedded into the sputtered U-Mo film and caused a nodular growth defect. (b) A U-Mo flake that has dropped onto a Al-6061 substrate prior to Zry-4 and U-Mo deposition. The flake was completely covered by a layer of Zry-4 and U-Mo.

Concept for improvement

We tried to avoid the wall pollution by a regular mechanical cleaning of the chamber walls. By this mean we could significantly reduce the particle dropping, but not completely avoid it, as every deposition run generates a new layer. The effect could however easily be avoided, if the deposition setup would be turned by 90° or 180°, so that the particles can not any longer fall onto the substrates but would fall away from the substrates. This simple mean was unfortunately not applicable to the tabletop reactor, due to a lack of space in the fume hood it is installed in. It can however be considered for a future reactor improvement.

5.3 Target material spectrum

Current status

For all applications shown in this thesis we operated our sputtering reactors only in DC mode and used a pure physical sputtering reaction. This is sufficient to process most of the relevant materials in the field of U-Mo fuel fabrication, which are mainly metals (Al, Zr, Ti, Ni, Ta, Bi, Cd), alloys (U-Mo, AlFeNi, Al-6061, Zry-4) or semiconductors (Si), that means electrically conductive materials. Advanced fuel designs require however the processing of non-conductive compounds like ZrN, which could be an effective diffusion preventive coating material [izh09], or B_4C, which could be used as potent neutron poison [kei11]. The processing of these materials is neither possible by pure physical sputtering nor by DC sputtering.

Concept for improvement

Two different methods can be used to deposit a chemical compound onto a substrate: first, a sputtering target of that particular compound can be used in a diode sputtering setup to deposit a film. If the compound is conductive, a DC diode is in principle sufficient for that purpose. Non-conductive target materials on the other side can be processed, if the diode is operated in OF mode. Second, the compound can be formed during the sputtering process. One or more elements of the compound are provided as ejected material flux from the sputtering target, the remaining ones are added with the working gas. The compound film on the substrate forms as a result of a chemical reaction of the different components.

In practice, usually both methods are used together. RF sputtering on the one side allows to use any material as a sputtering target, regardless if it is electrically conductive or not. Basically any solid material can be sputter eroded in a RF diode and thus any elemental composite of a solid target can be provided as sputtered particle flux. The working gas on the other side is prepared from reactive and non-reactive gaseous composites to provide elements that cannot be used as target materials, gases for example. The chemical reactions on the target surface, in the plasma state and on the substrate surface will then determine, how the sputtered deposit will be composed.

RF sputtering It has been mentioned already in section 2.2, that the DC diode shown in figure 2.11 can only be operated with conductive target materials. The surface of a non-conductive target would immediately accumulate positive charge due to the bombarding ions and by that build up an electric field, that would stop any further surface ion bombardment and also extinguish the glow discharge. A permanent inversion in the diode polarity can however prevent this, if the frequency of the pole reversal is high enough to avoid a surface charge accumulation. A frequency of about 50 - 100 kHz is usually sufficient to achieve that [ros90]. The diode has thus to be operated in OF mode to process both conductive and non-conductive materials. Usually a frequency of 13.56 MHz is used for the OF sputtering (in this case also called radio frequency or RF sputtering), as this frequency is officially assigned for this type of applications.

The process of ion bombardment in RF sputtering is different from DC sputtering. Both electrodes, target and substrate, will alternately be subject to ion bombardment in RF mode, but the intensity will be massively reduced compared to the DC case. This is because for pole reversal frequencies in the MHz range, the ions in the discharge plasma can hardly follow the electric field oscillations and therefore hardly move at all, as their electrical mobility[2] is very small. The electrons

---

[2]The electrical mobility μ is defined as the ability of a charged particle to move through a medium in response to an electric field E. The formal equation is $\mu = \frac{v}{E} = \frac{q}{m\nu}$, where v is the

5 ADVANCEMENT

on the other side have a comparably high mobility and follow the field easily. As the ion current at the cathode has however to be identical to the electron current at the anode, the low ion mobility reduces the electron as well as the ion bombardment of both electrodes.

To achieve an efficient target ion bombardment in RF operation, the DC diode setup has to be modified as shown in figure 5.12a [ros90]. The central insight in this respect is, that the discharge plasma together with the two electrodes can be seen as a serial connection of two capacitors C_{target} and $C_{substrate}$ (see figure 5.12b). For symmetric area electrodes, as in the DC case, C_{target} equals $C_{substrate}$ and the same voltage U builds up at each of the electrodes leading to a small and alternating electron and ion bombardment. If the electrodes are however asymmetric in area, C_{target} and $C_{substrate}$ are different and therefore different voltages U_{target} and $U_{substrate}$ will build up at the two electrodes, a behavior which is denoted as self biasing. The ratio between voltages and capacitors is given as $\frac{U_{target}}{U_{substrate}} = \frac{C_{substrate}}{C_{target}}$, so a higher negative voltage will build up at the smaller electrode while the larger one will generate a higher positive voltage (see figure 5.12c). If an additional coupling capacitor is inserted into the circuit, that prevents a reversal of current direction, the self bias effect will make the smaller electrode to a negatively charged 'RF cathode', while the larger electrode will become a positively charged 'RF anode'. A small target electrode will thus be the RF cathode and by that face an intense ion bombardment and sputter erosion, while a large substrate electrode will be the RF anode and face a strongly reduced ion bombardment (and sputter erosion).

By choosing the capacitors C_{target} and $C_{substrate}$ adequately, it is even possible to completely avoid the sputter erosion of the substrate electrode. The necessary condition is $q \cdot U_{substrate} <$ sputtering energy threshold, where q is the charge of the bombarding ions.

Reactive sputtering   Reactive sputtering is possible both in a DC or a RF operated sputtering diode and allows to deposit a multitude of chemical compounds. The available elements for the chemical reactions in the reactive sputtering process are constituted by the used target material and by the working gas composition. The process behavior and chemistry is however primarily determined by the processing parameters, especially by the flow of reactive working gas [dep08].

Figures 5.13a and b illustrate schematically the characteristic process behavior of a reactive sputtering process with a metallic target in dependency of the working gas flow. As it can be seen, both the sputter erosion rate (figure 5.13a) and the

---

drift velocity, m is the particle mass, q is the particle charge and ν is the collision frequency with particles of the medium.

Target material spectrum

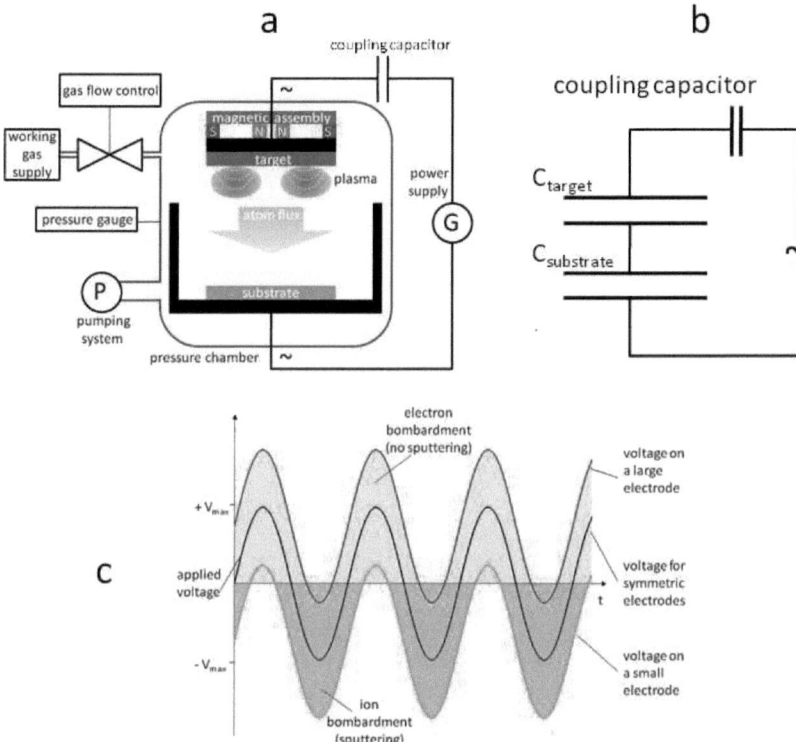

Figure 5.12: (a) Basic assembly of a diode sputtering setup for OF operation. The setup is nearly identical to the one shown in figure 2.11. The electrodes are however asymmetrically in size and a coupling capacitor has been added to the circuit. During OF operation this setup generates a negative self bias voltage at the target electrode. (b) Simplified circuit diagram of the OF setup. The discharge plasma together with the two electrodes can be seen as a serial connection of two capacitors C_{target} and $C_{substrate}$. The ratio between capacitor and the voltage at each capacitor is given as $\frac{U_{target}}{U_{substrate}} = \frac{C_{substrate}}{C_{target}}$. (c) The applied voltage from the OF generator is supposed to be of sine shape. If both electrodes would be identical in size, each would face the same ion and electron bombardments and by that be alternately sputter eroded or coated. If the sizes of the electrodes are not identical however, the different mobilities of electrons and ions cause a self biasing. The smaller electrode is preferentially bombarded by ions while the larger electrode will preferentially be bombarded by electrons. Thus in the OF case the relative size of the electrode determines, whether it is the sputtering target or substrate. If the substrate capacitance $C_{substrate}$ is large enough, the ions bombarding the substrate surface will have energies below the sputtering energy threshold and the substrate electrode won't be subject to sputter erosion any more.

149

5 ADVANCEMENT

reactive working gas pressure (figure 5.13b) show a hysteresis behavior if the gas flow is increased or decreased. The hysteresis separates two different phases of process behavior, that are sometimes denoted as 'metallic' phase and 'poisoned' phase [dep08], but are typical for most reactive sputtering processes also when non-metallic targets are used.

For low reactive gas flows (up to the position A in the two figures), the sputtering of target material will dominate reactive process behavior. Sputter erosion and deposition rates in this phase are relatively high compared to the latter one. The reactive gas flow is nearly completely consumed by chemical reactions at the substrate surface, in the plasma and at the target surface and the reactive gas partial pressure is thus low. The deposited films are usually under-stoichiometric in this phase, as the amount of target material flux exceeds the reactive gas supply. The target surface will continuously get partially covered by chemical compounds produced in the reactions, but sputter erosion will always remove these compounds and keep the surface mainly metallic. The sputtering diode can thus be operated both in DC or RF mode without problem. Increasing the reactive gas flow in this situation will lead to an increasing number of chemical reactions in the process.

A critical condition is reached (position A in the two figures), when the whole target surface is completely covered (or: 'poisoned') by a compound layer. As the sputtering yields of target material and compound material are usually very different, the sputter erosion, target material flux and deposition rates decrease drastically at this condition. Due to the reduced target material flux, the number of chemical reactions in the plasma and at the substrate surface also drops immediately, and not the whole amount of supplied reactive gas is consumed any more. As a consequence the reactive gas partial pressure rises quickly (position B in the two figures), and the excessive reactive gas supply usually leads to the deposition of over-stoichiometric films. A poisoned target can moreover only be operated in DC mode, if the compound layer on the target surface is electrically conductive. Otherwise it can only be operated in RF mode, as the DC operation would stop as soon as the target is poisoned.

The way back from a poisoned to a metallic target surface requires a reduction of reactive gas flow below the position B and continuous ion bombardment to remove the poisoned layer from the target surface (position C in the two figures, but reversed direction). Once the surface is substantially metallic (position D in the two figures), the starting state has been reached again.

The hysteresis effect limits the spectrum of achievable compositions of chemical compounds in a reactive sputtering process [scl11]. The film stoichiometries that are reached for those processing conditions, at which the hysteresis effect occurs, are usually hard to realize. It is however possible to shift the 'position' of the

Target material spectrum

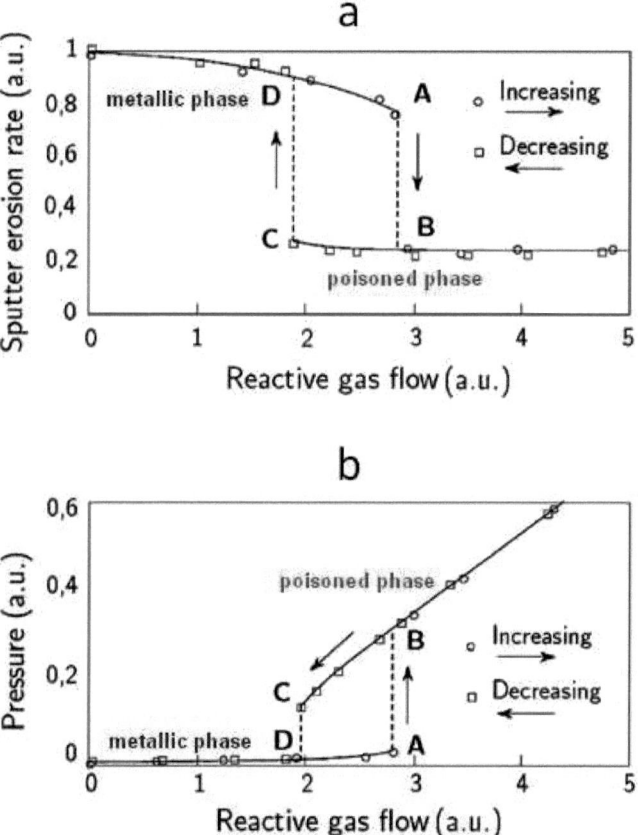

Figure 5.13: Figures from [dep08]: (a) Schematic sputter erosion rate of a metallic target in dependency of the reactive gas flow. (b) Schematic reactive gas partial pressure in dependency of the reactive gas flow. Both figures show the characteristic hysteresis behavior that is typical for reactive sputtering.

5 ADVANCEMENT

desired deposit	ZrN	B$_4$C
diode operation	DC or RF	RF
target material	Zr	B$_4$C
working gas	Ar + N$_2$	Ar
gas composition (vol %)	92 (Ar) 8 (N$_2$)	100 (Ar)
substrate temperature (°C)	350	300
bias voltage (V)	-350	+15

Table 5.1: Target material, working gas composition and processing parameters for the successful deposition of ZrN (from [re03]) and B$_4$C (from [pas98]).

hysteresis loop, that means to control the range of reactive gas flow at which the hysteresis effect appears, simply by adjusting the working gas pumping speed. A higher pumping speed allows to feed larger amounts of reactive gas into the sputtering process without poisoning the target.

In summary it can be stated, that the deposition of a particular chemical compound by reactive sputtering is unfortunately much more complex than in non-reactive sputtering, as not only target material and working gas composition have to be specified, but also the appropriate processing parameters have to be identified to actually deposit a compound with desired stoichiometry.

Compounds for fuel fabrication Currently two chemical compounds are relevant for U-Mo fuel processing, ZrN [izh09] and B$_4$C [kei11]. The deposition of both materials by sputter deposition has been studied for applications in microelectronics as well as for hard coatings. ZrN is usually produced by reactive sputtering using a Zr target and Ar/N$_2$ working gas in a sputtering diode operated either in DC or RF mode. B$_4$C is usually produced by non-reactive RF sputtering using a B$_4$C target and Ar working gas. Table 5.1 lists processing parameters, that have been reported to successfully produce the two deposits. The reactive gas flux is however not given, as it is dependent on the used pumping speed and thus on the design of the sputtering reactor.

5.4 Target utilization

Current status

The efficiency of the sputter processing in our reactors could be improved significantly, if the material utilization of the target would be increased. For the current setups it was determined to be around 20 - 25 % for the tabletop reactor respectively 30 - 35 % for the full size reactor, literature describes however sputtering techniques that allow a degree of utilization of more than 95 % [ang03].

Concept for improvement

Our sputtering reactors use a static assembly of sputtering target and magnets (schematically shown in figure 5.14a). Dependent on the size of this assembly, it is possible to eject up to 30 - 35 % of the total target mass until the sputtering target has to be replaced (see chapter 3.2.2). By the following methods it would however be possible to increase this percentage:

- The magnetic assembly, that defines the position of the discharge plasma ring, can be moved periodically and relatively to the target (see figure 5.14b). For an appropriate motion path this leads to a uniform target ion bombardment and erosion and thus to a high material utilization of >80 % [fre87]. This method has however the disadvantage, that suspension structures as the target mounting screws, will be eroded as well. To avoid this, these structures could be made from non-conductive materials (for DC operation) or grounded (for RF operation). The most common method is however a target suspension from the back side.

- The shape of the magnetic assembly can be changed to homogenize the magnetic field along the racetrack (see figure 5.14c). This measure (often referred to as 'magnetron balancing') is primarily intended to avoid the CCE, but it also increases material utilization moderately to up to 30 - 40 % [mat02].

- An extension of the field closing plates, as shown in figure 5.14d, allows to generate a magnetic field with field lines that are nearly parallel to the target surface. The resulting discharge plasma volume is adapted to the space between the field closing plates and thus leads to an erosion zone that can be adapted to the target geometry. This allows a material utilization of > 80 % [may97].

- The glow discharge can be operated at elevated pressures and without the magnetic assembly as a normal glow discharge (see figure 5.14e). This leads

5 ADVANCEMENT

to a uniform ion bombardment of the target and could in theory allow to achieve a material utilization of >80 %, but reduces the sputtering rate significantly. The normal glow discharge moreover requires pressures in the range of 10^{-2} - 10^1 mbar. As a consequence certain film growth conditions (regions in the Thornton zone model with pressures below 10^{-2} - 10^1 mbar) are not available any more for the deposition process.

- The plasma needed for ion bombardment can be generated by an external plasma generator (this principle is called 'ion beam sputtering', see figure 5.14f), that allows a controlled localized erosion of the target. Currently the so-called 'High Target Utilization Sputtering' process (or: 'HiTUS') is known to reach in this way the highest possible material utilization for solid targets of > 98 % [ang03].

- Even in the static assembly shown in figure 5.14a a higher degree of material utilization can be reached if the solid sputtering target is replaced by a liquid 'target pool' that continuously transports material into the erosion areas (see figure 5.14g). This design can in principle reach higher degrees of material utilization (> 99 % [kna93]) than all the previously shown designs, as it does not require target change but just a continuous material supply to the target pool. The material ejection from this pool will however be a mixture between sputtering and thermal evaporation, which makes the controlled deposition of multicomponent materials difficult.

Target utilization

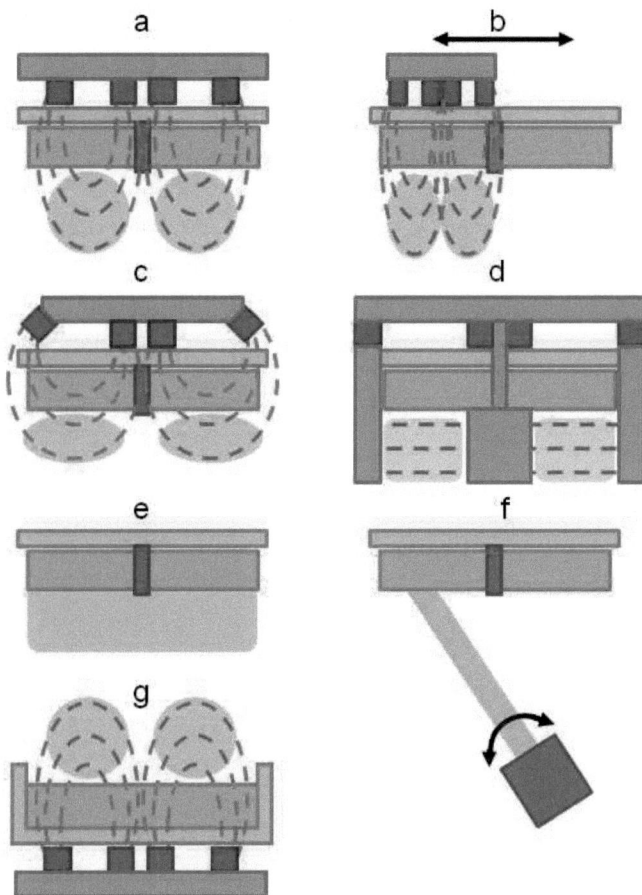

Figure 5.14: Different schematic target assemblies to improve target material utilization. (a) Setup with static magnetic assembly as it is currently used in our sputtering reactors. The degree of material utilization is 20 - 25 %. (b) Target setup with dynamic magnetic assembly that allows a material utilization of > 80 %. (c) Target setup with optimized magnetic geometry that avoids CCE and also increases material utilization moderately to 30 - 40 %. (d) Target setup with extended field closing plates that allow a material utilization of > 80 %. (e) Target setup operated with a normal glow discharge without magnetic enhancement. The degree of material utilization can be increased to > 80 %. The deposition rate is however decreased significantly and low pressure growth conditions are not available. (f) A plasma beam is used for a controlled erosion of the target surface and allows to reach a material utilization of > 98 %. (g) Target setup that is nearly identical to (a) but with a sputtering target consisting of a pool of liquid target material. This design allows to reach a material utilization of > 99 % but has the disadvantage, that material will continuously evaporate from the pool and produce a mixed flux of evaporated and sputtered material.

Chapter 6

Conclusion

This final chapter completes the presented thesis. The work that has been conducted is briefly summarized and conclusions are made.

6.1 Summary

The following engineering and scientific works have been performed in the context of this thesis:

- A tabletop sputtering reactor has been designed and constructed to conduct basic deposition experiments and to produce samples. It was installed into a fume hood in a radioisotope lab, and is in regular operation since several years now. It allows to quickly produce tailored samples for U-Mo metallurgy and irradiation experiments with little effort.

- A full size sputtering reactor has been designed and constructed to conduct sputter processing on U-Mo full sized foils. It was mounted in a glove box and installed into a hot lab facility. After an extensive technical inspection by the TÜV[1], the reactor was officially approved and received the operating license by the BStMUGV[2] as a permanent technical installation in the hot lab. It is in regular operation since one year.

- Both sputtering reactors have been studied and characterized in terms of process parameters as well as reactor, target and substrate properties. The sputtering process in these reactors is now well enough understood to allow the processing of U-Mo fuel foils and the fabrication of experimental samples without any difficulty.

---
[1] 'Technischer Überwachungs Verein' (english: 'German Technical Supervisory Association')
[2] 'Bayerisches Staatsministerium für Umwelt, Gesundheit und Verbraucherschutz' (english:'Bavarian State Ministry of the Environment, Public Health and Consumer Protection')

6 CONCLUSION

- The microscopic sputtering behaviour of the available target materials has been studied by simulations with the Monte Carlo program SRIM. The results of this simulations gave us a better understanding of the sputtering process.

- The deposition simulation program SPUSI has been developed. It has been used to study the macroscopic deposition behavior in our sputtering reactors, and to test means of optimization for them. The simulations performed by SPUSI allowed us to define a method to reduce thickness gradients in the deposited layers.

- The application of sputtering in the monolithic U-Mo fuel fabrication process was investigated. We could show, that sputter deposition can actually be used to produce monolithic U-Mo fuel foils and also to completely clad such foils. This is however a very extensive method that currently seems to be too unefficient to be actually used for production. The use of sputtering for the surface treatment of U-Mo fuel seems however to be a very promising application. Especially the fuel surface cleaning by sputter erosion as well as the deposition of functional materials on the cleaned surfaces allow to reach a degree of surface preparation, that cannot be reached by any of the presently used fabrication techniques. In this aspect the sputtering technique can thus be considered as clearly superior, and the integration of a sputter cleaning/coating step into the fabrication procedure seems to be reasonable and feasible.

- The application of sputtering to produce samples for different experimental purposes was investigated and several hundred samples were produced. We could show, that these sputter deposited samples can be used in HII to investigate IDL formation between U-Mo and Al, which opens the possibility to test measures against it. Leenears was able to study the solid state reaction between U and Si by using very thin U-Mo films that had been sputter deposited.

6.2 Conclusion

The application of sputter erosion and deposition in fuel plate fabrication appeared us to be a promising idea, but had not been investigated for monolithic U-Mo fuel up to this thesis. The main aim of this thesis thus was the realization of an experimental setup, that allowed to perform both sputter erosion and sputter deposition on full size U-Mo fuel foils. A further aim was to demonstrate and to study the process and its usability.

Conclusion

We constructed a sputtering reactor, that is suited to conduct sputter processing on U-Mo full sized foils. In experiments with this reactor we could show, that sputtering is a superior process for foil surface cleaning and functional layer application. Moreover sputtering allows to produce tailored samples for experimental studies making it a valuable tool in nuclear fuel research. Both applications are novel and promising, and will certainly find their place in U-Mo fabrication and research.

Appendix

A1: SRIM

The program SRIM ('Stopping and Range of Ions in Matter'), that was developed by Ziegler et al. [zie84], allows to simulate the interactions of ions with matter. We used the program to simulate the microscopic sputtering process, as well as to estimate the behavior of sputtered samples in heavy ion irradiation experiments.

A1.1: Program

SRIM is based on a Monte Carlo method and simulates the interaction of incident ions with the atoms of a plane and homogeneous material layer. Parameters like ion type, energy and angle of incidence as well as the type, density and thickness of material can be freely chosen as starting parameters. The program simulates a chosen number of ions including their trajectories, collisions and subsequent collision cascades in detail, and, based on that, calculates a multitude of output parameters including sputtering rate and energy distribution of sputtered particles. SRIM can therefore be used to simulate the microscopic sputtering process and to gain information on some of its essential parameters.
There are however several aspects that should be considered when using SRIM. First it should be mentioned, that SRIM uses the the binary collision approximation, which means that the influence of neighboring atoms on a single collision event is neglected. This means for sputtering in general, that only reactions in the single-collision and the linear cascade regime should be calculated by SRIM to get reasonable results. This is however not a drawback in our case, as we wanted to restrict our studies on this reactions (see chapter 2.1). Secondly, crystal structures and topological structures in general are neglected in the program. The bombarded material in the simulation is always a homogeneous, amorphous and flat material layer without any detailed surface or volume structure. Thus only the basic sputtering process for an ideal amorphous target will be calculated properly, the results for a target deviating from that may differ. A third aspect is, that SRIM does not account for dynamic effects such as composition changes or

6 CONCLUSION

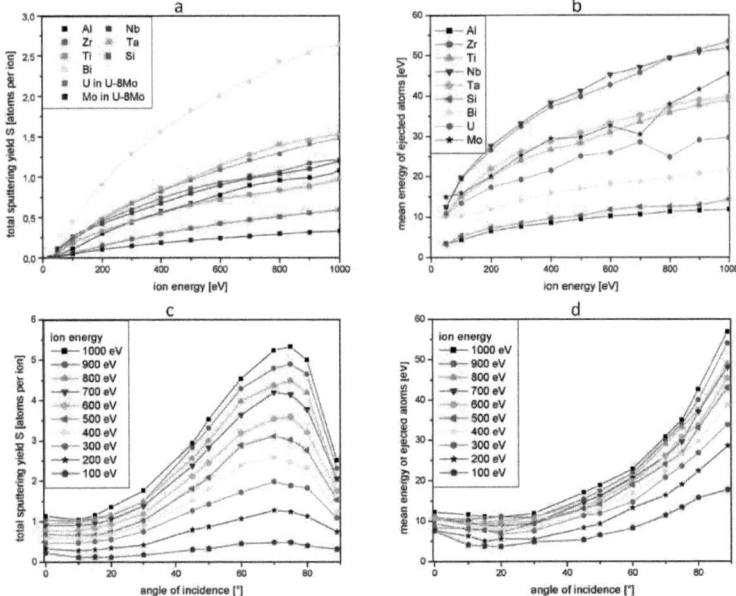

Figure 6.1: (a) Total sputtering yield per ion energy for the relevant materials calculated by SRIM. (b) Mean energy of ejected atoms for normal ion incidence depending on the ion energy. (c) Total sputtering yield per angle of incidence for different ion energies for Al. (d) Mean energy of ejected atoms per angle of incidence for different ion energies for Al.

evolution of radiation damages. A simulation of long-time effects is therefore not possible.

In summary, SRIM allows us to simulate and study the fundamental sputtering event itself within some limitations, and therefore it is a valuable tool. However it does not allow to simulate the complete sputtering process.

A1.2: Sputtering yield and mean energy

We used SRIM to calculate the total sputtering yield S per ion as a function of energy for normal ion incidence for all materials listed in table 3.4. We assumed a bombardment with single charged Ar ions and an acceleration voltage of 0 - 1 kV. Figure 6.1a shows the results. Apparently, the total sputtering yields for all materials stay between 0 - 2.5 atoms per ion for the given energy range. Figure 6.1b shows the mean energy of the ejected atoms, that are in the range of 5 - 50 eV.

We know from the previous chapter, that both the sputtering yield and the ejection energy are dependent from the angle of ion incidence. The principal behavior can be seen in figures 6.1c and d at the example of Al. It is clearly visible, that the maximum sputtering yield can be expected for angles between 70 - 75° to the surface normal. This range apparently represents an optimum between the size of the collision cascade and its proximity to the surface. A similar behavior can be expected for the other materials.

With the calculated sputtering yields S, it is in principle possible to calculate the ejection rates R in our reactors by the following equation:

$$R = \int_0^{U \cdot q \cdot e} \int_0^{90} S(E,\theta) \cdot I(E,\theta) d\theta dE$$

where U is the plasma voltage, q the maximum appearing charge state of bombarding ions, e the elementary charge, $S(E,\theta)$ the sputtering yield in dependence of ion energy and angle of ion incidence θ, and $I(E,\theta)$ the number of ions with a certain energy and a certain direction. The energetic and angular distribution $I(E,\theta)$ of the bombarding ions is usually hard to determine however. Thus we will be content by giving a rough estimate on R.

Assuming the conditions of the example for sputtering U-Mo, a plasma current of 1 A and single charged ions, we can roughly estimate that about $6.24 \cdot 10^{18}$ ions (the equivalent of 1 C) bombard the target per second. From the calculated composition of the altered layer and from the total sputtering yields of U and Mo, an average sputtering yield of 0.20 atoms per ion can be calculated. This gives a number of $1.25 \cdot 10^{18}$ atoms, that are ejected per second at 240 W, with a composition of 80% U and 20% Mo. This is equivalent to a mass of 0.44 mg U-8Mo, that gets eroded every second from the target and is ejected into the sputtering reactor. This is in good accordance to the experimentally determined average target mass loss of approximately 0.5 mg/s.

A1.3: Ion backscattering

As mentioned before, the substrate and the deposited film are subject to an energetic particle bombardment during sputtering. The bombarding particles are target atoms ejected during the sputtering process as well as energetic neutrals, that are generated during ion bombardment by neutralization and reflection of ions from the sputtering target. The rate of bombardment can be estimated by SRIM.

The number of target atoms that bombard the deposited film is equivalent to the just calculated ejection rate R times the fraction of target atoms, that actually reach the deposited film. This fraction is strongly dependent on the size of the film and

6 CONCLUSION

material	Ar ion reflection probability
Zr	0.13
Ti	0.01
Nb	0.12
Bi	0.28
Si	0.00
Al	0.00
Zry-4	0.13
U-8Mo	0.27
Al-6061	0.00
AlFeNi	0.00

Table 6.1: Ar ion reflection probabilities for the materials listed in 3.4.

on the solid angle it covers seen from the target. This can be measured quite easily. For our reactors, we measured that about 15 - 25 % of the ejected atoms reach the deposited substrates (see chapter 4.2.1). Thus the rate of substrate bombardment should be in the range of $1.9 - 3.1 \cdot 10^{17}$ target atoms per second for the U-8Mo example.

The number of neutrals, that bombard the sputtered film, is hard to determine. Basically every bombarding ion, that gets reflected from the target surface, can become a bombarding neutral, if it is neutralized during collision. The probability of neutralization is however not known to us. The number of reflected ions can however be seen as an upper maximum of the number of bombarding neutrals. We thus calculated the reflection probability for all our target materials by SRIM (see table 6.1).

As SRIM uses the binary collision approximation, the reflection probability is independent of the collision energy. For our U-8Mo example, the reflection probability accounts 0.27, meaning that 27 % of all bombarding Ar ions are reflected from the target[3]. If all of this ions would be neutralized during collision, the sputter deposited film would face a bombardment of $2.5 - 4.2 \cdot 10^{17}$ neutrals per second.

A1.4: Atom reflection

The reflection of atoms from the surface of a solid appears, if the mass of the surface atoms is larger than the mass of the impinging atoms. For sputtering substrates, this effect usually states no problem at all, as the substrate surface

---
[3]This reflection is already accounted in the sputtering yield S.

Conclusion

material	reflection probability
Zry-4	0.01(Zr), 0.00(Sn)
U-8Mo	0.00(U), 0.13(Mo)
Al-6061	0.01(Al), 0.01(Mg)
AlFeNi	0.01(Al), 0.00(Fe), 0.00(Ni), 0.01(Mg)

Table 6.2: Reflection probabilities for the atoms of the most important constituents for the multi-component target materials listed in table 3.3.

gets quickly covered by a film and does not play a role any more for reflection[4]. For mono-component films this effect is also not of importance, as basically no reflection occurs if an atom of one certain element is impinging onto a film of the same element. Atom reflection thus only plays a role for the deposition of multi-component materials.

We calculated the reflection probability for all multicomponent materials listed in table 3.3 by SRIM (see table 6.2).

Only in the case of the Mo constituent in U-8Mo a significant amount of atom reflection can be seen. According to the simulation, a fraction of 13 % of all Mo atoms reaching the U-8Mo film surface get reflected from it. This leads to a systematically lowered Mo content in a sputtered deposit. A U-8Mo target thus produces a U-7Mo film during sputter deposition. This shift in composition is however constant and not dependent on processing parameters, and can therefore be accounted already during target selection. The deposition of an U-8Mo film by sputtering would hence require an U-9.2Mo sputtering target.

---

[4]The deposition rates in our sputtering reactors have been measured to be in the range of $Å/s$, thus the substrate surface will completely be covered by a film within several seconds.

6 CONCLUSION

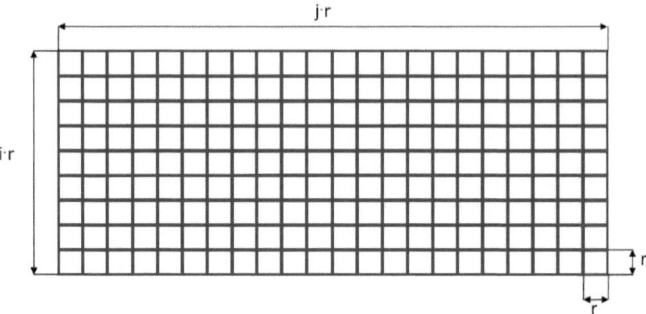

Figure 6.2: Target and substrate are modeled as rectangular lattices with a resolution r and a size i·r×j·r.

A2: SPUSI

The program SPUSI ('SPUttering SImulation') was developed by our technical student Bogenberger for the sputtering project. SPUSI allows to simulate the macroscopic deposition characteristic of a sputtering diode by just considering the geometry of the diode and simple linear transformations.

A2.1: Basic program

The basic model of SPUSI consists of a sputtering target and a substrate. The raw target is a rectangular lattice with resolution r and size i·r×j·r, that both can be preselected (see figure 6.2). The raw substrate is a rectangular lattice which is in size and resolution identical to the raw target.
In formal terms the target is defined as a matrix

$$t := t_{i,j}, \; i,j \in \mathbb{N}$$

and the substrate is a matrix

$$s := s_{k,l}, \; k,l \in \mathbb{N}, \; dim(s) = i \times j$$

where the single matrix elements are equivalent to target respectively substrate surface elements of size r×r. The numeric value of each target matrix element $t_{i,j}$ determines the value of sputter emission from that element, the value of a substrate matrix element determines the amount of deposited material on that element. In the raw target all elements are set to to the value 1, i.e. $t_{i,j} := 1 \; \forall \; i,j$, in the raw substrate all elements are set to the value 0, i.e. $s_{i,j} := 0 \; \forall \; i,j$.

Conclusion

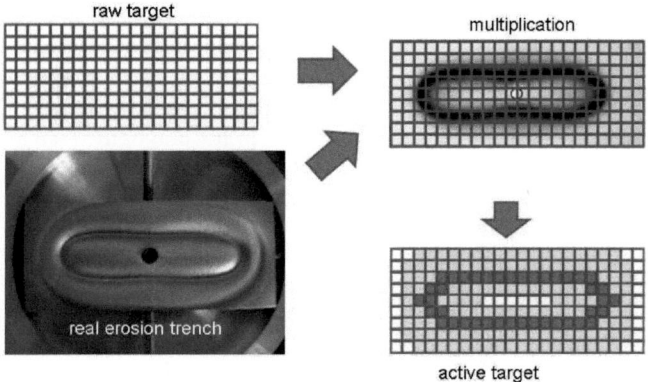

Figure 6.3: The raw target matrix is multiplied with a profile matrix of the actual erosion zone, which can be gained by a depth profile measurement. The result is an active target matrix. Different erosion rates at different positions are comprised within this active target as weighting factors of the single lattice elements (here symbolized by different shades of grey).

To account for the plasma position and the resulting erosion areas on the actual sputtering target in stationary configuration, the raw target matrix is multiplied with a profile matrix $p_{i,j} \in [0,1]$. It represents the three dimensional depth profile of the erosion trench. This profile matrix can easily be gained from a profile measurement, conducted for example with the ATOS system at the TUM Institute for Forming and Founding (see figures 3.17 and 3.18). The size of the matrix is determined by the size of the target in units of r. The matrix values are the normalized local depth values, where 1 is the maximum depth and 0 is the minimum depth. The value of each profile matrix element thus determines, how much erosion is visible in this element and, by that, how much material must have been ejected from that element. The result of this multiplication is the active target, that has active sputtering areas similar to an actual target (see figure 6.3). The corresponding matrix is defined as:

$$t_{i,j}^{act} := t_{i,j} \cdot p_{i,j}$$

In a next step the deposition process itself is modeled. We assume, that every single active target element continously ejects material with a certain angular distribution. This macroscopic angular distribution is determined by various microscopic effects and should not be mistaken with the microscopic angular distributions shown in figures 2.2 and 2.3. It results from the superposition of all microscopic ejection reactions inside the active target element, and thus contains all energetic, topographic and material influences listed in chapter 2.1. The macroscopic angular distribution function is not known, but has to be measured for each element, each target geometry and each energy. We also expect this distri-

6 CONCLUSION

bution function to change during operation, as the target geometry changes. As a working assumption we started with a macroscopic angular distribution function $\cos^x(\alpha)$, where α is the angle between the target surface normal and the direction of atom ejection, and x is a matching coefficient.

The angular distribution function defines, which fraction of the total amount of sputtered atoms is emitted in a certain direction. By applying it to a matrix element of the active target, one gains a rough model of the macroscopic sputter emission profile of this single target element. It is formally given by

$$t_{i,j}^{emi}(\alpha) := t_{i,j}^{act} \cdot \cos^x(\alpha), \; \alpha \in [0, \frac{\pi}{2}].$$

To gain the resulting deposition profile of this target element on the substrate, the projection of $t_{i,j}^{emi}(\alpha)$ on all substrate elements $s_{k,l}$ is calculated by the relation

$$s_{k,l} = \sum_i \sum_j t_{i,j}^{emi}(\alpha(k,l))$$

with

$$\alpha(k,l) = \arctan\left(\frac{\sqrt{(i_{target} - k_{substrate})^2 + (j_{target} - l_{substrate})^2}}{d}\right)$$

where d is the vertical distance between target and substrate in units of r, and $\sqrt{(i_t - k_s)^2 + (j_t - l_s)^2}$ is the horizontal distance between the target element $t_{i,j}$ and the substrate element $s_{k,l}$ in units of r (see figure 6.4).

The principal method is illustrated in figure 6.5.

A2.2: Implementation of movement

A possible way to reduce gradients that appear in the static configuration of SPUSI is the implementation of movement. It is possible to move either target or substrate or just the magnetic assembly (and thus the plasma itself). Every movement can however be described as an relative movement of the discharge plasma (material source) against the substrate. This relative movement was implemented into the basic SPUSI program by using the superposition of static sources and renormalizing.

The movement data is provided to the program as a list of target respectively substrate positions at certain[5] points of time $(x(t_i), y(t_i)) = (x_i, y_i), i \in \mathbb{N}$, that can

---
[5]Because intermediate coordinates are linearly interpolated it is reasonable to choose the given points accordingly.

Conclusion

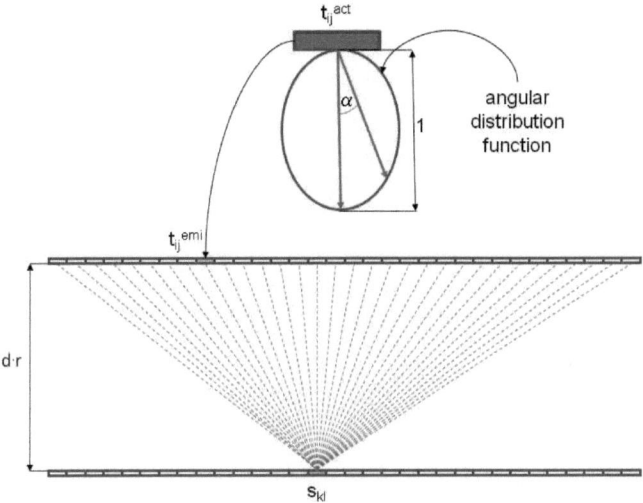

Figure 6.4: The deposition at the substrate matrix element $s_{k,l}$ is the sum of the emission of all active target elements $t_{i,j}^{emi}$ directed at the particular substrate matrix element $s_{k,l}$. The angular distribution function defines, which fraction of the total amount of sputtered atoms is emitted in a certain direction.

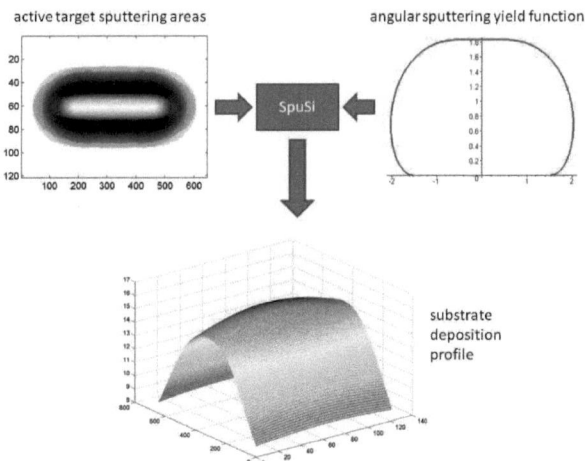

Figure 6.5: SPUSI needs the active target areas defined by the plasma as well as the angular distribution of the sputtered atoms to calculate the resulting deposition profile.

169

6 CONCLUSION

easily be created by a standard spreadsheet program. This given coordinates define a total approximate[6] path length D, along which the target/substrate will be moved:

$$D = \sum_{i=1}^{n-1} \sqrt{\left((x_i - x_{i+1})^2 + (y_i - y_{i+1})^2\right)}$$

The number n of discrete positions during this calculation is determined by the resolution in x- and y-direction, r_x and r_y:

$$n = \left\lceil \frac{D}{\sqrt{r_x^2 + r_y^2}} \right\rceil$$

The elements of the active target matrix $t_{i,j}^{act}$, which represent the active sputtering areas in a stationary configuration, are shifted according to the listed positions and superponed with each other. The intermediate result is an effective active target matrix

$$t_{i,j}^{act,eff,moved} = \frac{\sum_{v=1}^{n} t_{i(v),j(v)}^{act,eff}}{n}$$

as it is shown in figure 6.6, that is static but contains the movement in its structure. The deposition can hereafter be modeled like a stationary deposition process in the basic SPUSI program.

A2.3: Implementation of masks

As another option to reduce gradients we considered the use of a mask between target and substrate. The mask was implemented into SPUSI as a mask matrix $m_{q,r}$ that has the same size and resolution as $t_{i,j}$. The distance d between target and substrate is divided into two parts by the mask:

$$d = d_{target \to mask} + d_{mask \to substrate}$$

The mask matrix $m_{q,r}$ defines the permeability for the linear particle trajectories starting at the target. All elements of $m_{q,r}$ have either the numerical value 0 (meaning no transmission) or 1 (meaning full transmission). The mask is implemented into the sputtering process via a factor $\delta(q,r)$. We define:

$$\delta(q,r) = \begin{cases} 0, & \text{if } \alpha_{target/mask} = \alpha_{target/substrate} \\ 1, & \text{else} \end{cases}$$

---
[6] Again, this is owed to the fact that the path is interpolated linearly.

Conclusion

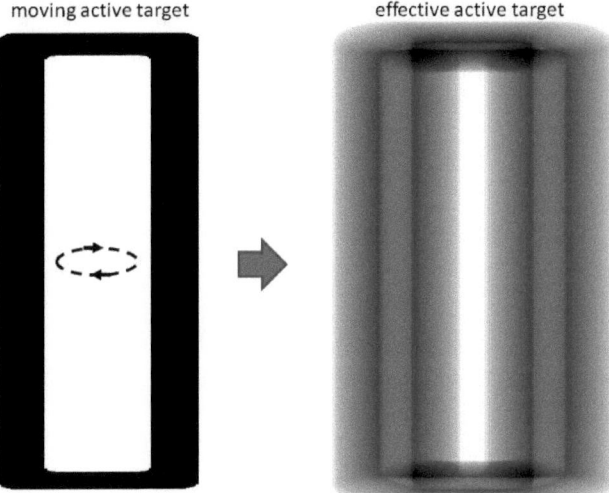

Figure 6.6: SPUSI displays the dynamic deposition process as a static process. In experiment the sputtering target (left side) should be moved according to the elliptic path. SPUSI shifts the active target area according to the path and superposes the resulting images to gain an effective active target area (right side). This static target contains the complete movement information but can be treated like in stationary deposition.

6 CONCLUSION

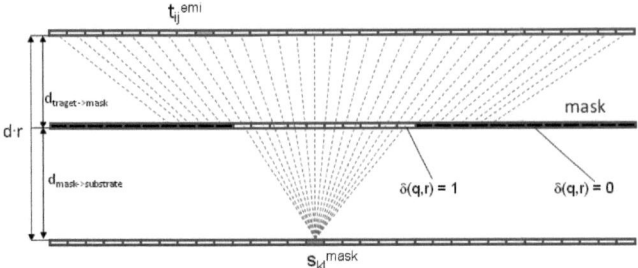

Figure 6.7: Mask concept. For every single substrate element $s_{k,l}^{mask}$ the deposition from any target element $t_{i,j}^{emi}(\alpha(k,l))$ is calculated. The mask represents a criteria of weighting. Some directions are weighted with 1 (no change) others with 0 (erasement). For every combination of one target and one substrate element the whole mask matrix has to be calculated however.

with

$$\alpha_{target/mask} = \alpha(q,r) = \arctan\left(\frac{\sqrt{(i_{target}-q_{mask})^2 + (j_{target}-r_{mask})^2}}{d_{target \to mask}}\right)$$

and

$$\alpha_{target/substrate} = \alpha(k,l)$$

The deposition profile during deposition with mask thus is given as:

$$s_{k,l}^{mask} = \sum_i \sum_j \sum_q \sum_r \delta(q,r) \cdot t_{i,j}^{emi}(\alpha(k,l))$$

Figure 6.7 illustrates the concept.

Conclusion

A3: Metallurgy of U-Mo

Pure metallic U is the material with the highest possible U density of 19.06 $\frac{gU}{cm^3}$ at room temperature. An application of pure metallic U as nuclear fuel is however hardly possible, as the material would be subject to anisotropic geometry changes due to phase changes under in-pile conditions [hol58].

A3.1: Metallic U

Jungwirth [jun11] states, that pure metallic U undergoes three phase changes between room temperature and its melting point at 1132 °C.
The room temperature phase of U is the orthorhombic α-phase (see figure 6.8a). It is stable up to 667 °C and shows a very large and also anisotropic thermal expansion behavior[7]. This behavior can result in undesired geometry changes of objects fabricated from α-U, which excludes it as a fuel material.
In the temperature range between 667 and 775 °C the U β-phase is stable (see figure 6.8b). It shows a tetragonal lattice structure and also an anisotropic thermal expansion behavior[8], which means it is also not suited as a nuclear fuel material.
The body-centered cubic U γ-phase (see figure 6.8c) is stable above 775 °C and shows an isotropic thermal expansion behavior[9]. It is thus the desired phase for U based fuel materials.
The U γ-phase is a high temperature phase, that decays at room temperature into the low temperature phases after some time and has thus to be stabilized. The alloying of γ-U with transition metals has shown to be an effective measure for phase stabilization, especially with Mo by forming U-Mo alloys.

A3.2: U-Mo alloys

As already mentioned in chapter 1.2.3, U-Mo alloys were identified as materials with high U density that also show an excellent irradiation performance up to high burnups. The term 'excellent irradiation performance' refers to the stability of the γ-U-Mo phase during in-pile irradiation, and to the retention of gaseous fission products inside the alloy matrix, which is given by their solubility in U-Mo.
The Mo content in the U-Mo determines the degree of γ-phase stability. A Mo content of 4.5 - 15.5 wt% has been determined to be able to stabilize the γ-phase at room temperature [jun11] (see phase diagram in figure 6.9). In this given

---
[7] $\frac{\Delta x}{x} = 35.6 \cdot 10^{-6}/°C$, $\frac{\Delta y}{y} = -8.4 \cdot 10^{-6}/°C$, $\frac{\Delta z}{z} = 31.6 \cdot 10^{-6}/°C$
[8] $\frac{\Delta x}{x} = 23.6 \cdot 10^{-6}/°C = \frac{\Delta y}{y}$, $\frac{\Delta z}{z} = 10.4 \cdot 10^{-6}/°C$
[9] $\frac{\Delta x}{x} = 21.6 \cdot 10^{-6}/°C = \frac{\Delta y}{y} = \frac{\Delta z}{z}$

6 CONCLUSION

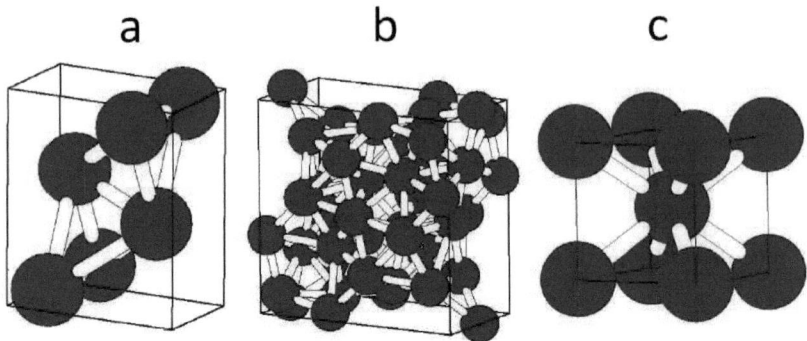

Figure 6.8: Figures from [jun11]: (a)The orthorhombic U α-phase is stable up to 667 °C. Due to its very large anisotropic thermal expansion behavior it is not possible to use U based fuel materials containing this phase, as the fuel might possibly undergo massive deformation during reactor operation. (b) The tetragonal U β-phase is stable in the temperature range 667 - 775 °C. It also shows anisotropic thermal expansion behavior, which also disqualifies it for utilization. (c) The body-centered cubic U γ-phase is stable above 775 °C and shows an isotropic thermal expansion behavior. It is thus the desired phase for U based fuel materials.

range, a larger Mo content stabilizes the γ-phase also for elevated temperatures (see figure 6.9), while a smaller Mo content increases the U density of the U-Mo alloy. In practice, the exact composition of a U-Mo alloy has to be adapted to the particular requirements and applications it will be used for. If the alloy undergoes a thermal treatment during fabrication - for example if hot rolling techniques are used - the Mo content has to be chosen adequately to allow the γ-phase to withstand the treatment.

The phase stability and transformation behavior of a particular U-Mo alloy is illustrated by the Time-Temperature-Transformation (TTT) diagrams (see figure 6.10). It should be noted, that the TTT diagrams only cover thermal phase transformations, not irradiation induced ones. This is of special importance, as during in-pile irradiation α-phase U-Mo and U_2Mo is known to transform into γ-phase U-Mo regardless of temperature [jun11].

Conclusion

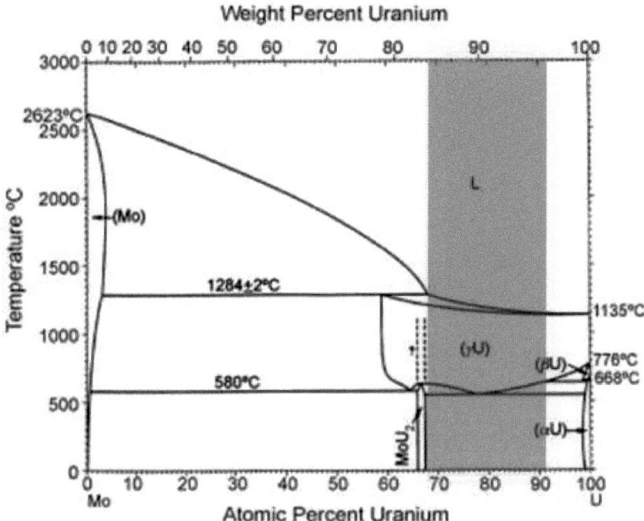

Figure 6.9: Phase diagram of the U-Mo system. Only the alloys with a Mo content of 4.5 - 15.5 wt% (grey zone) stabilize the U γ-phase and are thus relevant as fuel meterials.

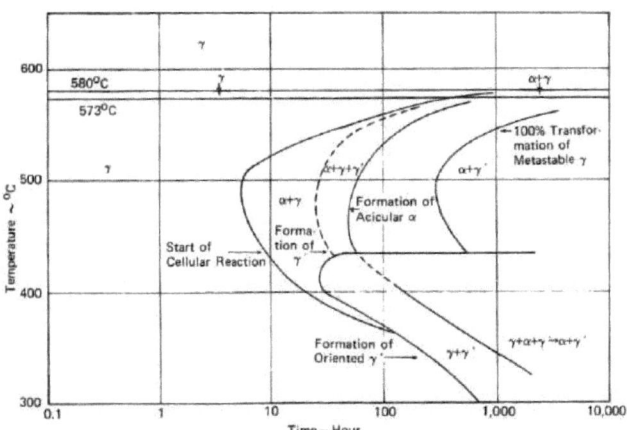

Figure 6.10: Time-Temperature-Transformation (TTT) diagram for the alloy U-10wt%Mo, that illustrates the decomposition kinetics of the γ-phase (from [jun11]). The critical temperature of U-10Mo is 580°C. Above this temperature the γ-phase is permanently stable, below it is only metastable, that means it decomposes into other phases over time. The decomposition mechanisms are usually complex and involve several distorted phases (as γ').

175

6 CONCLUSION

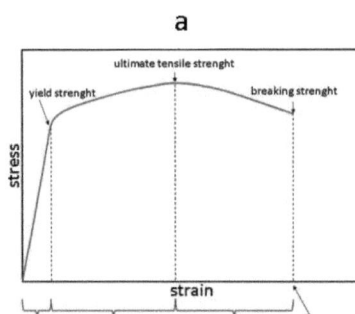 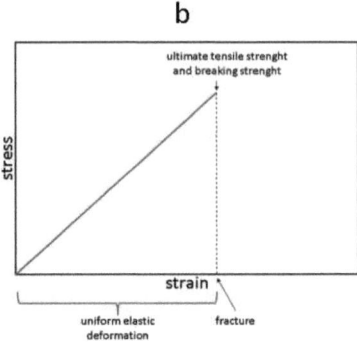

Figure 6.11: (a) Schematic of a typical stress-strain curve for a ductile tensile test sample. In the first region of elastic deformation, the stress rises proportional to the strain. This is called the 'Hook area'. When the yield strength is reached, the sample is deformed inelastically. A small rise in stress causes the strain to increase over proportional. The maximum stress value that can be applied is called the 'ultimate tensile strength'. After this value has exceeded, the sample can be further deformed by application of an even decreasing value stress until the breaking strength is reached and a fracture occurs. (b) Schematic of a typical stress-strain curve for a non-ductile tensile test sample. The test stays always in the region of elastic deformation. Fracture occurs, if the point for inelastic deformation is reached. For the samples we use for our adhesion measurements the behavior is quite similar. Fracture can however occur even far before inelastic deformation will start.

A4: Tensile tests

Tensile tests are usually prepared on wires, strips or machined samples of a material to measure its tensile strength. The samples are screwed into jaws or gripped in them, and stretched by moving the grips apart at a constant rate while measuring the applied load and the grip separation. The tensile strength is the stress at which the material breaks or deforms permanently. Three different types of tensile strength are usually distinguished: yield strength, ultimate tensile strength and breaking strength (see figure 6.11a).

The yield strength denotes the tensile strength value, at which the elastic deformation behavior switches to a plastic deformation. The ultimate tensile strength denotes the tensile strength value, at which the maximum stress value on the stress-strain curve appears. The breaking strength is finally the tensile strength value at the point of fracture.

A tensile test to measure the adhesion of different materials to each other is a special case of the described general tensile test. The sample geometry has to be adequate to actually measure the adhesion of the materials, not their particular tensile strength. Usually the materials to test have a planar interface to each other and the tensile force is applied perpendicular to this interface. The adhesion of the different materials in such a sample is generally equal or smaller than the breaking strength of the weakest material involved. Usually the adhesion is however even below the yield strength of the weakest material, as the beginning of plastic deformation in one part of the sample often causes the materials to loose the adhesion to each other. Tensile test samples to investigate the adhesion of different materials thus show very often the tensile behavior of non-ductile materials (see figure 6.11b). The breaking strength in this case can be identified with the strength of the material adhesion.

Bibliography

[ang03] J.R. Anguita, *HiTUS Technology as presented to Institute of Physics*, Plasma Quest Limited, 2003

[ari10] S. Aricó, S. Balart, A. Bonini, P. Cristini, L. de Lio, L. Dell'Occhio, D. Gil, A.G. Gonzalez, R. González, C. Komar Varela, M. López, M. Mirandou, H. Taboada, *2010 National progress report on R&D on LEU fuel and target technology in Argentina*, 32nd International Meeting on Reduced Enrichment for Research and Test Reactors (RERTR), Lisbon, 2010

[beh81] R. Behrisch, P. Sigmund, M. T. Robinson, H. H. Andersen, H. L. Bay, H. E. Roosendaal, *Sputtering by Particle Bombardment I: Physical Sputtering of Single-Element Solids*, Springer Verlag, Garching, 1981

[beh83] R. Behrisch, G. Betz, C. Carter, B. Navinsek, J. Roth, B. M. U. Scherzer, P.D. Townsend, G. K. Wehner, J. L. Whitton, *Sputtering by Particle Bombardment II: Sputtering alloys and compounds, electron and neutron sputtering, surface topography*, Springer Verlag, Garching, 1983

[beh91] R. Behrisch, K. Wittmaack, W. Hauffe, W. O. Hofer, N. Laegreid, E. D. McClanahan, B. U. R. Sundqvist, M. L. Yu, *Sputtering by Particle Bombardment III: Characteristics of sputtered particles, technical applications*, Springer Verlag, Garching, 1991

[bir06] G.A. Birzhevoi, A.D. Karpin, V.V. Popov, V.N. Sugonyaev, *Some approaches to solving the problem of diminishing the interaction between U-Mo fuel particles and Al matrix*, 10th International Topical Meeting on Research Reactor Fuel Management, Sofia, 2006

[bir09] G.A. Birzhevoi, A.D. Karpin, V.V. Popov, V.N. Sugonyaev, *Methods of increasing the uranium charge in fuel elements of research reactors*, 13th International Topical Meeting on Research Reactor Fuel Management, Vienna, 2009

[boe82] K. Böning, W. Gläser, R. Golub, J. Meier, *The proposed spallation neutron source and modernized reactor as possible sites for a low temperature irradiation*

BIBLIOGRAPHY

facility in Germany, Journal of Nuclear Materials, Volumes 108-109, Pages 10-20, Garching, 1982

[boe98] K. Böning, *Richtfest - Neue Forschungsneutronenquelle Garching*, Informational booklet, Technische Universität München, Munich, 1998

[boe02] K. Böning, *Use of highly enriched uranium at the FRM II*, 6th International Topical Meeting on Research Reactor Fuel Management, Ghent, 2002

[boe04] K. Böning, W. Petry, A. Röhrmoser, Chr. Morkel, *Conversion of the FRM II*, 8th International Topical Meeting on Research Reactor Fuel Management, Munich, 2004

[bre11] H. Breitkreutz, *Neutronics and thermal hydraulics of high density cores at FRM II*, Doctoral thesis, Technische Universität München, Munich, 2011

[bun94] Bunshah, *Handbook of deposition technologies for films and coatings (Second edition)*, ISBN 0-8155-1337-2, 1994

[cah94] R.W. Cahn, P. Haasen, E.J. Kramer, *Nuclear materials part I*, Materials science and technology: a comprehensive treatment Vol. 10 A, VHC Verlagsgesellschaft, Weinheim, 1994

[can10] L. Canella, *Optimization of the PGAA instrument at FRM II for low background and 2D measurement*, Doctoral thesis, Technische Universität München, Munich, 2010

[can11] L. Canella, Wolfgang Schmid, Winfried Petry, Andreas Türler, *Composition analysis of U-Mo sputtering targets and magnetron sputtered U-Mo films by two dimensional PGAA*, Paper submitted to the Journal of Nuclear Materials, Munich, 2011

[che06] ChenYang Technologies GmbH + Co.KG, *Neodymium Iron Boron magnets Catalogue*, Finsing, 2006

[cla03] C.R. Clark, G.C. Knighton, M.K. Meyer, G.L. Hofman, *Monolithic fuel plate development at Argonne National Laboratory*, 25th International Meeting on Reduced Enrichment for Research and Test Reactors (RERTR), Chicago, 2003

[cla06] C.R. Clark, J.F. Jue, G.A. Moore, N.P. Hallinan, B.H. Park, *Update on Monolithic Fuel Fabrication Method*, 28th International Meeting on Reduced Enrichment for Research and Test Reactors (RERTR), Cape Town, 2006

[cle05] *Spent fuel reprocessing: a fully mastered pathway*, CLEFS CEA Nr. 53, 2005

BIBLIOGRAPHY

[dep08] D. Depla, S. Mahieu *Reactive sputter deposition*, Springer-Verlag Berlin Heidelberg, Gent, 2008

[dir10] S. Dirndorfer, *Tensile tests on monolithic two metal layer systems for research reactor fuels*, Diploma thesis, Technical University Munich, Garching, 2010

[dub06] S. Dubois, H. Palancher, F. Mazaudier, P. Martin, C. Sabathier, M. Ripert, P. Lemoine, C. Jarousse, M. Grasse, N. Wieschalla, W. Petry, *Development of UMo/Al dispersion fuel: an oxide layer as a protective barrier around the UMo particles*, 28th International Meeting on Reduced Enrichment for Research and Test Reactors (RERTR), Cape Town, 2006

[edi02] T.A. Edison, *Process of coating phonograph records*, United states patent Nr.713863, Llewellyn Park, 1902

[fan03] Q.H. Fan, L.Q. Zhou, J.J. Grácio, *A cross-corner effect in a rectangular sputtering magnetron*, Journal of Physics D: Applied Physics 36, p. 244, 2003

[for09] *Forschung mit Neutronen - Methoden entwickeln, Natur befragen, Antworten bekommen*, Informational booklet, Technische Universität München, Garching, 2009

[fre87] H. Frey, G. Kienel *Dünnschichttechnologie*, VDI Verlag GmbH, Düsseldorf, 1987

[fue57] Various authors, *Fuel elements conference, Book 1*, Paris, 1957

[gla99] W. Gläser, *Plans for the utilization of the new research reactor FRM II*, Technical University Munich, Garching, 1999

[ham05] J.M. Hamy, P. Lemoine, F. Huet, C. Jarousse, J.L. Emin, *The french U-Mo group contribution to new LEU fuel development*, 9th International Topical Meeting on Research Reactor Fuel Management, Budapest, 2005

[han96] N.A. Hanan, S.C. Mo, R.S. Smith, J.E. Matos, *An alternative LEU design for the FRM II*, 18th International Meeting on Reduced Enrichment for Research and Test Reactors (RERTR), Seoul, 1996

[har04] G. Harbonnier, P. Colomb, JP. Durand, B. Duban, Y. Lavastre, *Success story of FRM-II fuel element manufacturing*, 8th International Topical Meeting on Research Reactor Fuel Management, Munich, 2004

[hen08] R. M. Hengstler, *Thermal and electric conductivity of a monolithic uranium-molybdenum alloy for research reactor fuels*, Diploma thesis, Technical University Munich, Garching, 2008

BIBLIOGRAPHY

[hen09] R. M. Hengstler, L. Beck, H. Breitkreutz, C. Jarousse, R. Jungwirth, W. Petry, W. Schmid, J. Schneider, N. Wieschalla, *Physical Properties of Monolithic U-8wt%Mo*, Journal of Nuclear Materials, Erlangen, 2009

[her04] Wolfgang A. Hermann, *Neutronen sind Licht: Die Forschungs-Neutronenquelle Heinz Maier-Leibnitz in Garching*, Plenarvortrag Jahrestagung Kerntechnik, Düsseldorf, 2004

[hof87] G.L. Hofman, L.A. Neimark, *Prospects for stable high-density dispersion fuels*, Argonne National Laboratory, Argonne, 1987

[hof99] G.L. Hofman, M.K. Meyer, J.L. Snelgrove, M.L. Dietz, R.V. Strain, K-H. Kim, *Initial assessment of radiation behavior of very-high-density low-enriched-uranium fuels*, 22nd International Meeting on Reduced Enrichment for Research and Test Reactors (RERTR), Budapest, 1999

[hof04] G.L. Hofman, M.R. Finlay, Y.S. Kim, *Postirradiation analysis of low enriched U-Mo/Al dispersion fuel miniplate tests, RERTR-4 and -5*, 26th International Meeting on Reduced Enrichment for Research and Test Reactors (RERTR), Vienna, 2004

[hof06] G.L. Hofman, Y.S. Kim, H.J. Ryu, J. Rest, D.M. Wachs, M.R. Finlay, *Attempt to solve the instability in the irradiation behavior of low enriched U-Mo/Al dispersion fuel*, 10th International Topical Meeting on Research Reactor Fuel Management, Sofia, 2006

[hol58] A.N. Holden, *Physical metallurgy of uranium*, Vallecitos Atomic Laboratory, General Electric Company, Pleasanton, California, Atoms for Peace, Genf, 1958

[izh09] A.L. Izhutov, V. Alexandrov, A. Novosyolov, V. Starkov, A. Sheldyakov, V. Shishin, V. Yakovlev, I. Dobrikova, A. Vatulin, V. Suprun, Ye. Kartashov, V. Lukichev, *The status of LEU U-Mo fuel investigation in the MIR reactor*, 31st International Meeting on Reduced Enrichment for Research and Test Reactors (RERTR), Beijing, 2009

[jar09] C. Jarousse, L. Halle, W. Petry, R. Jungwirth, A. Röhrmoser, W. Schmid *FRM II and AREVA-CERCA common effort on monolithic UMo plate production*, 13th International Topical Meeting on Research Reactor Fuel Management, Vienna, 2009

[joh04] R.E. Johnson, R.W. Carlson, J.F. Cooper, C. Paranicas, M.H. Moore, M.C. Wong, *Radiation effects on the surfaces of the Galilean satellites* Jupiter. The planet, satellites and magnetosphere, Cambridge, 2004

BIBLIOGRAPHY

[jun06] R. Jungwirth, *Thermische und elektrische Leitfähigkeit von hochdichten Uran-Molybdän-Kernbrennstoffen*, Diploma thesis, Technical University Munich, Garching, 2006

[jun10] R. Jungwirth, W. Petry, H. Breitkreutz, W. Schmid, H. Palancher, C. Sabathier, *Study of heavy ion irradiated U-Mo/Al miniplates: Si and Bi addition to Al and U-Mo ground powders*, 14th International Topical Meeting on Research Reactor Fuel Management, Marrakech, 2010

[jun11] R. Jungwirth, *Irradiation behavior of modified high-performance nuclear fuels*, PhD thesis, Technical University Munich, Garching, 2011

[jur10] H. Juranowitsch, *Cross section examination of sputtered uranium-molybdenum nuclear fuels*, Diploma thesis, Technical University Munich, Garching, 2010

[jur11] H. Juranowitsch, W. Schmid, R. Jungwirth, S. Dirndorfer, W. Petry, *Characterization of bond strenght of inverse monolithic two metal layer*, Publication in preparation, Garching, 2011

[kei11] D. Keiser et al., *Microstructural characterization of burnable absorber materials being evaluated for application in LEU U-Mo fuel plates*, 15th International Topical Meeting on Research Reactor Fuel Management, Rome, 2011

[kim05] Y.S. Kim, G.L. Hofman, H.J. Ryu, J. Rest, *Thermodynamic and metallurgical considerations to stabilizing the interaction layers of U-Mo/Al dispersion fuel*, 27th International Meeting on Reduced Enrichment for Research and Test Reactors (RERTR), Boston, 2005

[kna93] K.E. Knapp, *Method and apparatus for sputtering of a liquid*, United States Patent 5211824, 1993

[lee10] A. Leenaers, S. Van den Berghe, *In-situ x-ray diffraction study of the U(Mo)/Si solid state reaction*, 32nd International Meeting on Reduced Enrichment for Research and Test Reactors (RERTR), Lisboa, 2010

[lie05] M.A. Lieberman, A.J. Lichtenberg, *Principles of plasma discharges and materials processing*, John Wiley and Sons, Berkeley, 2005

[lop02] A. Lopp, C. Braatz, M. Geisler, H. Claus, J. Trube, *Plasma simulation for planar sputtering cathodes*, 45th Annual Technical Conference Proceedings of the Society of Vacuum Coaters, Lake Buena Vista, 2002

[mat02] Materials Science Inc., *Magnetron sputtering*, Technology note, San Diego, 2002

BIBLIOGRAPHY

[mat03] D.M. Mattox, *The foundations of vacuum coating technology*, Noyes Publications, Albuquerque, 2003

[may97] R. Mayer, *Optimierung eines DC-Sputtermagnetrons*, Doctoral thesis, Universität Kaiserslautern, Kaiserslautern, 1997

[mey00] M.K. Meyer, G.L. Hofman, R.V. Strain, C.R. Clark, J.R. Stuart, *Metallographic Analysis of Irradiated RERTR-3 Fuel Test Specimens*, 23rd International Meeting on Reduced Enrichment for Research and Test Reactors (RERTR), Las Vegas, 2000

[mey02] M.K. Meyer, G.L. Hofman, S.L. Hayes, C.R. Clark, T.C. Wiencek, J.L. Snelgrove, R.V. Strain, K.-H. Kim, *Low-temperature irradiation behavior of uranium-molybdenum alloy dispersion fuel*, Journal of Nuclear Materials 304, pages 221-236, Argonne, 2002

[mo89] S.C. Mo, J.E. Matos, *Conversion feasibility studies for the Grenoble High Flux Reactor*, 12th International Meeting on Reduced Enrichment for Research and Test Reactors (RERTR), Berlin, 1989

[mo95] S.C. Mo, N.A. Hanan, J.E. Matos, *Comparison of the FRM-II HEU design with an alternative LEU design* and *Attachment to comparison of the FRM-II HEU design with an alternative LEU design*, 17th International Meeting on Reduced Enrichment for Research and Test Reactors (RERTR), Paris, 1995

[moo08] G.A. Moore, F.J. Rice, N.E. Woolstenhulme, W.D. Swank, D.C. Haggard, J. Jue, B.H. Park, S.E. Steffler, N.P. Hallinan, M.D. Chapple, D.E. Burkes, *Monolithic fuel fabrication process development at the Idaho National Laboratory*, 30th International Meeting on Reduced Enrichment for Research and Test Reactors (RERTR), Washington, 2008

[moo10] G.A. Moore, J-F. Jue, B.H. Rabin, M.J. Nilles, *Full size U-10Mo monolithic fuel foil and fuel plate fabrication-technology development*, 14th International Topical Meeting on Research Reactor Fuel Management, Marrakech, 2010

[mov69] B.A. Movchan, A.V. Demchishin, *Investigations of the structure and properties of thick Ni, Ti, W, Al2O3 and ZrO2 vacuum condensates*, Fizika Metalov i Metalovedenije, 28, No.4, 1969

[nor10] figures by Kai Nordlund, professor of computational materials physics, University of Helsinki

[nud00] M. Nuding, M. Rottmann, A. Axmann, K. Böning, *FRM II project status and safety of its compact fuel element*, 4th International Topical Meeting on Research Reactor Fuel Management, Colmar, 2000

BIBLIOGRAPHY

[ohr02] M. Ohring, *Materials science of thin films: deposition and structure*, Second Edition, Academic Press, New Jersey, 2002

[pal06] H. Palancher, P. Martin, C. Sabathier, S. Dubois, C. Valot, N. Wieschalla, A. Röhrmoser, W. Petry, C. Jarousse, *Heavy ion irradiation as a method to discriminate research reactor fuels*, 10th International Topical Meeting on Research Reactor Fuel Management, Sofia, 2006

[par05] J.M. Park, H.J. Ryu, G.G. Lee, H.S. Kim, Y.S. Lee, C.K. Kim, Y.S. Kim, G.L. Hofman, *Phase stability and diffusion characteristics of U-Mo-X (X=Si, Al, Zr) alloys*, 27th International Meeting on Reduced Enrichment for Research and Test Reactors (RERTR), Boston, 2005

[pas04] E.E. Pasqualini, M. López, *Increasing the performance of U-Mo fuels*, 26th International Meeting on Reduced Enrichment for Research and Test Reactors (RERTR), Vienna, 2004

[pas98] E. Pascual, E. Martínez, J. Esteve, A. Lousa, *Boron carbide thin films deposited by tuned-substrate RF magnetron sputtering*, Diamond and Related Materials 8 (1999), p.402-405, 1999

[pat86] K. C. Radford, *Coating a nuclear fuel with a burnable poison*, United States Patent 4587088, 1986

[pat91] W.J. Bryan, N. Fuhrman, D.C. Jones, *Element with burnable poison coating*, United States Patent 4990303, 1991

[pat93] W.J. Bryan, D.C. Jones, *Wear resistant coating for fuel cladding*, United States Patent 5268946, 1993

[pat06] P. Böni, N. Wieschalla, *Method for producing a fuel element for a nuclear reactor*, International patent WO/2007/059851 2006

[per09] E. Perez, B. Yao, Y.H. Sohn, D.D. Keiser Jr., *Interdiffusion in Diffusion Couples: U-Mo vs. Al and Al-Si*, 31st International Meeting on Reduced Enrichment for Research and Test Reactors (RERTR), Beijing, 2009

[pri10] Private communication with Walter Carli, chief operator of the tandem accelerator at the Maier-Leibnitz Laboratory in Garching, Garching, 2010

[que98] Y. Quéré, *Physics of materials*, Overseas publishers association, Amsterdam, 1998

[rai91] Y.P. Raizer, *Gas discharge physics*, Springer Verlag, Heidelberg, 1991

BIBLIOGRAPHY

[rap09] *Rapport transparence et sécurité nucléaire 2009*, Réacteur Haut Flux - Institut Laue Langevin, Grenoble, 2009

[re03] M. Del Re, R. Gouttebaron, J.-P. Dauchot, P. Leclère, G. Terwagne, M. Hecq, *Study of ZrN layers deposited by reactive magnetron sputtering*, Surface and Coatings Technology 174-175, p. 240-245 2003

[rertr] Homepage of the RERTR program, *www.rertr.anl.gov*

[rob09] A. B. Robinson, G. S. Chang, D. D. Keiser Jr., D. M. Wachs, D. L. Porter, *Irradiation performance of U-Mo alloy based 'monolithic' plate-type fuel - design selection* Idaho National Laboratory, Idaho Falls, 2009

[rob10] A.B. Robinson, D.M. Wachs, G.S. Chang, M.A. Lillo, *Summary of 'AFIP' full sized plate irradiations in the advanced test reactor*, 14th International Topical Meeting on Research Reactor Fuel Management, Marrakech, 2010

[roe05] A. Röhrmoser, W. Petry, N. Wieschalla, *Reduced enrichment program for the FRM-II, Status 2004/05*, 9th International Topical Meeting on Research Reactor Fuel Management, Budapest, 2005

[ros90] S.M. Rossnagel, J.J. Cuomo, W.D. Westwood, *Handbook of plasma processing technology: fundamentals, etching, deposition and surface interactions*, Noyes Publications, New Jersey, 1990

[rip09] M. Ripert, F. Charollais, M. C. Anselmet, X. Tiratay, P. Lemoine, *Results of the IRIS4 irradiation in OSIRIS Reactor*, 31st International Meeting on Reduced Enrichment for Research and Test Reactors (RERTR), Beijing, 2009

[sch11] W. Schmid, S. Dirndorfer, R. Jungwirth, H. Juranowitsch, W. Petry, T. Zweifel, *Tailored model systems for IDL investigation and diffusion barrier optimization*, 15th International Topical Meeting on Research Reactor Fuel Management, Rome, 2011

[scl11] S M. Schleussner, *ZrN Back-Contact Reflectors and Ga Gradients in $Cu(In,Ga)Se_2$ Solar Cells*, PhD thesis, University of Upsala, Upsala, 2011

[shi00] E. Shidoji, M. Nemoto, T. Nomura, *An anomalous erosion of a rectangular magnetron system*, Journal of Vacuum Science and Technology A, 18, p. 2858, 2000

[sne95] J.L. Snelgrove, G.L. Hofman, T.C. Wiencek, C.T. Wu, G.F. Vandegrift, S. Aase, B.A. Buchholz, D.J. Dong, R.A. Leonard, B. Srinivasan, D. Wu, A. Suripto, Z. Aliluddin, *Development and processing of LEU targets for Mo-99 production - Overview of the ANL program*, 18th International Meeting on Reduced Enrichment for Research and Test Reactors (RERTR), Paris, 1995

[sne96] J.L. Snelgrove, G.L. Hofman, C.L. Trybus, T.C. Wiencek, *Development of very-high-density fuels by the RERTR Program*, 19th International Meeting on Reduced Enrichment for Research and Test Reactors (RERTR), Seoul, 1996

[sne99] J.L. Snelgrove, G.L. Hofman, M.K. Meyer, S.L. Hayes, T.C. Wiencek, R.V. Strain, *Progress in developing very-high-density low-enriched-uranium fuels*, 3rd International Topical Meeting on Research Reactor Fuel Management, Bruges, 1999

[sta07] W.M. Stacey *Nuclear reactor physics I, Second edition*, WILEY-VCH Verlag GmbH & Co. KGaA, Weinheim, 2007

[ste01] M. Stepanova, S.K. Dew, *Estimates of differential sputtering yields for deposition applications*, Journal of vacuum science and technology, Vol. 19, Issue 6, page 2805, Alberta, 2001

[ste11] C. Steyer, *Sputter coating of spherical powder*, Diploma thesis, Technical University Munich, Garching, to be published

[tan61] K. Tangri, G.I. Williams, *Metastable phases in the uranium-molybdenum system and their origin*, Journal of Nuclear Materials 4, Nr.2, p.:226-233, Amsterdam, 1961

[tec08] D.M. Wachs, C.R. Clark, R.J. Dunavant, *Conceptual process description for the manufacture of low-enriched uranium-molybdenum fuel*, Technical report, Idaho National Laboratory, Idaho Falls, 2008

[tg3] Bayrisches Staatsministerium für Landesentwicklung und Umweltfragen, *Teilgenehmigung nach §7 Atomgesetz (AtG) zum Betrieb der Hochflussneutronenquelle München in Garching (FRM-II) - 3. Teilgenehmigung*, München, 2003

[tho74] J.A. Thornton, *Influence of apparatus geometry and deposition conditions on the structure and topography of thick sputtered coatings*, Journal of Vacuum Science and Technology, Vol. 11, 1974

[tho77] J.A. Thornton, *High rate thick film growth*, Annual reviews of material science 1977.7: 239-260, 1977

[van10] S. van den Berghe, A. Leenaers, C. Detavernier, *Selenium fuel : surface engineering of U(Mo) particles to optimise fuel performance*, 14th International Topical Meeting on Research Reactor Fuel Management, Marrakech, 2010

[wac08] D.M. Wachs et al., *Progress in US LEU fuel development*, 12th International Topical Meeting on Research Reactor Fuel Management, Hamburg, 2008

BIBLIOGRAPHY

[was92] K. Wasa, S. Hayakawa, *Handbook of sputter deposition technology: principles, technology and applications*, Noyes Publications, Osaka, 1992

[wie06] N. Wieschalla, *Heavy ion irradiation of U-Mo/Al dispersion fuel*, Technical University Munich, Garching, 2006

[xou04] N. Xoubi, R.T. Primm III, *Modeling of the High Flux Isotope Reactor cycle 400*, ORNL technical report, Oak Ridge, 2004

[zie84] J.F. Ziegler, J.P. Biersack, U. Littmark, *The stopping and range of ions in solids*, Vol. 1 of the series 'Stopping and Ranges of Ions in Matter', Pergamon Press, New York, 1984

[zho98] Zhou L., Wei X., *Formation of nodular defects as revealed by simulation of a modified ballistic model of depositional growth*, Journal of Materials Science, 33, p.1487 - 1490, 1998

Die VDM Verlagsservicegesellschaft sucht für wissenschaftliche Verlage abgeschlossene und herausragende

Dissertationen, Habilitationen, Diplomarbeiten, Master Theses, Magisterarbeiten usw.

für die kostenlose Publikation als Fachbuch.

Sie verfügen über eine Arbeit, die hohen inhaltlichen und formalen Ansprüchen genügt, und haben Interesse an einer honorarvergüteten Publikation?

Dann senden Sie bitte erste Informationen über sich und Ihre Arbeit per Email an *info@vdm-vsg.de*.

Sie erhalten kurzfristig unser Feedback!

VDM Verlagsservicegesellschaft mbH
Dudweiler Landstr. 99
D - 66123 Saarbrücken

Telefon +49 681 3720 174
Fax +49 681 3720 1749

www.vdm-vsg.de

Die VDM Verlagsservicegesellschaft mbH vertritt

Printed by Books on Demand GmbH, Norderstedt / Germany